Getting the Message

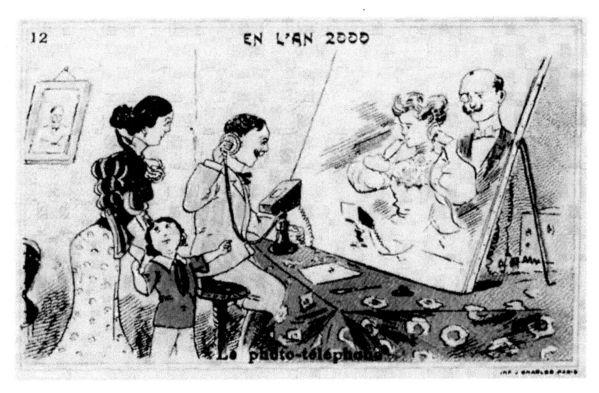

A nineteenth-century prediction of the state of the art in the year 2000.

Getting the Message

A History of Communications

Second Edition

Laszlo Solymar

OXFORD
UNIVERSITY PRESS

OXFORD
UNIVERSITY PRESS

Great Clarendon Street, Oxford, OX2 6DP,
United Kingdom

Oxford University Press is a department of the University of Oxford.
It furthers the University's objective of excellence in research, scholarship,
and education by publishing worldwide. Oxford is a registered trade mark of
Oxford University Press in the UK and in certain other countries

First edition published in 1999
Second edition published in 2021

Impression: 1

Published in the United States of America by Oxford University Press
198 Madison Avenue, New York, NY 10016, United States of America

British Library Cataloguing in Publication Data
Data available

Library of Congress Control Number: 2021932103

ISBN 978–0–19–886300–7
DOI: 10.1093/oso/9780198863007.001.0001

Printed and bound by
CPI Group (UK) Ltd, Croydon, CR0 4YY

To my grandchildren Juliet, Oscar,
Georgina, and Tanya

Preface to the Second Edition

The history of communications is a branch of the history of technology. However, most of technology delivers something tangible: a piece of machinery, a piece of furniture, a road, a bridge, or a plastic bag, to name a few. The goods produced by communications are quite different. They are messages: nearly always useless but occasionally very useful. They were already used at the dawn of civilisation for early warning, for receiving information about approaching armies.

The first edition, published in 1999, was mainly about point-to-point communications as realized by the telegraph (mechanical or electrical), the telephone, the fax machine, the telex, microwave links, satellite and optical communications. I excluded broadcasting whether radio or television. I made though a concession, by describing a kind of broadcasting by telephone that was founded in 1893 in Budapest, and even survived the First World War. I did include the fledgling Internet and made some predictions about the future.

In the new edition, as in the old one, I start with some correspondence some 4,000 years ago between the King of Mari (a city on the banks of the Tigris) and whoever was in charge of communications at the time. The historical context is always emphasized, e.g. the Kruger telegram that caused the cooling of relations between Germany and the UK prior to the First World War, or the Ems telegram that led to the 1870 war between France and Germany. Social history related to communications like scandals, murder, and bankruptcy has also been included as much as genius, inventiveness, and steady progress.

Twenty-one years is a long time and particularly in communications that is probably the fastest advancing discipline. There has been enormous expansion in satellite communications, and similarly in optical communications, which jointly cover by now every corner of the Earth.

And there is the Internet, in its infancy at the time, that has turned into an aggressive and robust adult. I am going to discuss both the advantages (Internet is so all-embracing that it is difficult to imagine life without it) and the abuses which are numerous. The same applies to smartphones, and particularly to the younger generation. I have a few photographs in Chapter 20 showing their obsession.

A notable inclusion into the new edition is the story of the Soviet InterNyet. It shows the difficulty of a dictatorship to cope with a technique that cannot be easily controlled. They never managed to set up an all-embracing computer network.

A discipline that is still in its infancy is artificial intelligence. It is included in a separate chapter in order to discuss its potential. Are the claims advanced by those doing research in the subject sustainable? Finally, I add a chapter on the future. Technical advances are quite predictable but otherwise (peace, politics, society) predictions are risky and it is not easy to be optimistic.

Acknowledgements First Edition

First of all I wish to acknowledge my debt to the Oxford College system which permits, nay, encourages the contacts between the practitioners of the arts and of the sciences. I had the good luck to be able to discuss Mesopotamia with Stephanie Dailey, the Holy Scriptures with John Barton, Classics with Stephanie West, Byzantine times with Philip Pattenden (actually, from Cambridge), science in the seventeenth century with Scott Mandelbrot, Napoleonic times with Geoffrey Ellis, the nineteenth and early twentieth centuries with Bob Evans, citations from Goethe with Kevin Hilliard, a translation from Confucius with Z. Cui, post-Second World War politics with Nigel Gould-Davies, the standard of living indices with Charles Feinstein, and matters in economics with Roger van Noorden and Tony Courakis.

Concerning the history of communications I wish to acknowledge the help I received from Patrice Carré and Christine Duchesne-Reboul of France Telecom; John Bray, Peter Cochrane, David Hay, Neil Johannesen, and H. Lyons of British Telecom; Alan Roblou of the BBC; Karoly Geher of the Technical University of Budapest; Tony Karbowiak of the University of New South Wales; Peter Kirstein of University College, London; David Payne of the University of Southampton; Victor Kalinin of Oxford Brookes University; and Dominic O'Brien, David Dew-Hughes, Terry Jones, Lionel Tarassenko, Don Walsh, and David Witt of the Department of Engineering Science, University of Oxford.

For help with the literature search I wish to thank Stephen Barlay, Leon Freris, Margaret Gowing, George Lawrence, Gabriella Netting, Sandor Polgar, Klaus Ringhofer, and Jeno Takacs. I am indebted to Michael Allaby, Eric Ash, Frank Ball, Mike Brady, Godfrey Hodgson, Gillian Lacey-Solymar, Avril Lethbridge, Lucy Solymar, and David Witt for reading various parts of the manuscript and for helpful comments. The whole of the manuscript was read and a large number of stimulating comments were made by Jonathan Coopersmith, Richard Lawrence, Julia Tompson, and Peter Walker.

I am greatly indebted to David Clark, the Head of the Department of Engineering Science, and Chris Scotcher, who is in charge of administration, for providing generous facilities while this book was written.

Special thanks are due to Jeff Hecht who let me read the manuscript of his book *The City of Light*, to Geoffrey Wilson who let me use any material from his book *The Old Telegraph*, and to Mark Neill for providing the pixellated pictures of Napoleon in Chapter 15. I have to mention separately Pierre-Louis Dougniaux, the picture archivist of France Telecom, who went well beyond the line of duty in providing me with photographs from the history of communications in France.

Finally, I wish to acknowledge the great debt I owe to my wife, Marianne, who not only read the manuscript but was willing to put up with the long hours I spent in libraries and archives.

Oxford   L.S.
October 1998

Acknowledgements Second Edition

I wish to thank John Holt for help on optical communications, Eric Ash and Richard Syms for discussions on the subject of communications in general, Ekaterina Shamonina for help with both text and drawings, and Alexander Shamonin for inside information on social media. Finally, I wish to thank my wife, Marianne, who encouraged me to write this second edition.

Oxford 2020

Contents

Figure Acknowledgements
First Edition

AT&T Fig. 9.4

Bodleian Library, University of Oxford Figs 3.6 (M90.D00642, p. 297), 4.12 (N2706.d10, 14/9/1850, p. 117), 4.13 (N2706.d10, 21/8/1858, p. 77), 5.3 (N2706.d10, 2/4/1892, p. 163), 5.4(b) (N2706.d10, 12/1/1910, p. 27), 6.9(a) (N2706.d10, 22/10/1913, p. 341), 7.1 (N2706.d10, 28/3/1891, p. 151), 7.2 (N2706.d10, 12/3/1913, p. 203), 22.1 (ALM2706d99, 1879)

British Telecom Archives Figs 4.7 (P4027, *c.* 1856), 4.11(a) (YB42, 1882), 5.4(a) (E73245, 1910), 5.9 (E6325, 1929), 7.6 (ARC14, c. 1907), 7.7 (Post84/8 Selection of publicity material, c. 1905)

Cable & Wireless Figs 6.1, 14.1

Corvina Kiado, Budapest Fig. 5.21

France Telecom, Archives et Documentation Historique Figs 2.3, 3.1, 4.4, 4.20, 5.5, 5.6, 5.7, 6.9(b), 7.5, 7.8, 9.11, 15.7, 16.6, 20.2

GloCall Satellite Services Fig. 9.13

Mark Harden Fig. 15.1(a)

Piers Helm Figs 3.12, 5.2, 16.1

Hertford College, Oxford Fig. 5.11

Illustrated London News 4.11(b), 15.6

Institution of Electrical and Electronic Engineers, New York Fig. 15.8

Intel Corporation/*Physics Today* Fig. 10.1

Marconi Electronic Systems Ltd Fig. 6.7

Mike Mosedale Fig. 14.6

Musée des Arts et Métiers, Paris Fig. 15.5(b)

Museum für Post und Kommunikation, Frankfurt am Main Fig. 5.8

National Gallery, London Fig. 1.1

Northern Electric plc, London Fig. 9.2

David Payne Figs 12.1, 12.8

Murray Ramsey Fig. 12.4

Geoffrey Wilson Figs 3.2(a), 3.5, 3.8

Acknowledgements for Figures Added in the Second Edition

G. P. Agrawal 18.1, 18.2, 18.4

Bankmycell 20.3

Carcharoth 9.10

Cartoon Collections 21.2, 21.3

Computer Museum Moscow 16.8(a)

FTTH Council of Europe 18.3

Getty Images 16.7, 21.1

History of Computing in Ukraine 16.8(b)

Peter Kirstein 16.5

Pew Research Centre 20.5, Table 20.1

The Planetary Society 9.7

NASA Earth Observatory 9.8

Statista 17.1

Union of Concerned Scientists 17.2

Shutterstock 20.7(a,b,c)

We are Social 20.6

Wikipedia 19.2

Wikimedia Commons 16.4

World Bank 20.1, 20.2

Part I
The First Thirty-Six Centuries

CHAPTER
ONE

Introduction

The history of communications is a branch of the history of technology but, strictly speaking, it is in a category of its own. The goods produced by technology, whether a piece of machinery, a piece of clothing, or a piece of furniture, are tangible; they perform some useful function. The goods produced by communications are messages. They are mostly useless but when they are useful they can be very, very useful. For that reason communication has always been regarded as a good thing by all peoples at all times. Even in prehistoric times a tribal chief would have easily appreciated both the military and economic implications. He would have dearly loved to receive reports like 'Scores of heavily armed Mugurus sighted at edge of Dark Dense Forest' or 'Buffalo herd fording Little Creek at Mossy Green Meadow'.

The idea was there but the means of sending messages were rather limited until very recent times. The same limitation did not apply to human imagination. A god in Greek mythology could contact any of his fellow gods without much bother and could cover the distance from Mount Olympus to, say, the battlefields of Troy in no time at all. Communication between gods was, of course, not possible in monotheistic religions. On the other hand the single god could easily send messages to any chosen individual. A possible method was first to call attention to impending communications (e.g. by a burning bush) and then deliver messages in a clear, loud voice. Oral communication was nearly always the preferred method but there is also an example of coded written communications in the Book of Daniel. The occasion is a feast given by Belshazzar, King of Babylon. Belshazzar draws upon himself the wrath of Jehovah by drinking with his wives and concubines from the holy vessels plundered earlier from the Temple in Jerusalem. Thereupon a message appears on the wall, MENE, MENE, TEKEL, UPHARSIN. This message is decoded by Daniel, as saying: 'God has numbered thy kingdom and finished it. Thou art weighed in the balances, and art found wanting'. By next morning Belshazzar was dead. This unique example of instantaneous written communications may be seen in Fig. 1.1 in Rembrandt's interpretation.

Besides appealing to human imagination, communications have a number of other distinguishing features. Its rate of progress over the past century and a half has been conspicuously faster[1] than that of any other human activity, and shows no sign of letting up. Let me make a few comparisons. In 1858 it took 40 days for the news of the Indian Mutiny to reach London.[2] By 1870 there were several telegraph lines

[1]This claim may be rightfully challenged by computer enthusiasts but it will be discussed in Chapter 16. Communications and computers are no longer separate subjects.
[2]To be exact, to reach Trieste, because by that time there was a telegraph connection between Trieste and London.

Getting the Message: A History of Communications. Second Edition. Laszlo Solymar, Oxford University Press (2021).
© Laszlo Solymar. DOI: 10.1093/oso/9780198863007.003.0001

Fig. 1.1 *Belshazzar's Feast* by Rembrandt.

connecting India to Europe. Transmission time depended mainly on the speed of re-transmission from station to station, four hours being a good estimate. The progress in 12 years from 1000 hours down to 4 hours represents an improvement by a factor of 250. For the Atlantic route the advent of the submarine cable in 1866 reduced the time for sending a message from a couple of weeks to practically instantaneous transmission. The figures are no less daunting if we talk about the capacity of a single line of communications then and now. In the 1840s when electrical telegraphy started to become widespread, information could be sent at a rate of about 4 or 5 words per minute. Today, the full content of the *Encyclopaedia Britannica* could be transmitted on a single strand of optical fibre in a fraction of a second. A similar increase in, say, shipping capacity would mean that a single ship would now be capable of transporting trillion tons of goods, i.e. more than a thousand tons for every man, woman, and child on Earth.

A third possible measure is the cost of information, not when we send information in bulk—that is less tangible—but when we want a leisurely chat with a friend in America. In 1927, when the trans-Atlantic telephone service was opened (relying on radio waves), a three-minute telephone call cost £15. Today, it might cost 10p. In nominal prices the reduction is by a factor of 30 which, in comparison to the figures quoted previously, is perhaps less striking, but since we are talking about money in our pocket, its effect on everyday life is much more significant. It needs to be added of course that prices have risen considerably since 1927. A loaf of bread, for example, cost about 3d. (1.25p in decimal currency) at the time, whereas today it costs something like £1. So while the price of bread has gone *up* by a factor of 80, the price of a trans-Atlantic telephone call has gone *down* by a factor of 150. In real terms, to make that call is now cheaper by an amazing factor of 12,000. And this is not an anomaly. We would arrive at similar figures whichever aspect of communications is chosen for comparison. The benefits are obvious. In 1927 only the richest people could afford a social telephone call across the Atlantic; today it is within the reach of practically all people in Europe or America.

What else is so extraordinary about communications? Its significance for conducting affairs of state. Governments which were quite happy leaving the manufacture of guns and battleships in private hands were determined to keep communications under their control. Perhaps the most forward-looking one was the French Government. As early as 1837, before the appearance of the electric telegraph, the Parliament approved the proposal that

Anyone who transmits any signals without authorization from one point to another one whether with the aid of mechanical telegraphs or by any other means will be subject to imprisonment for a duration of between one month and one year

This law was repealed only in the 1980s when France, following cautiously the example set by the United Kingdom, started on its privatization programme.

Having made a case for communications being a subject worthy of study, I would like to add that there is no chance whatsoever of doing it justice in a single book. Of necessity the subject must be restricted. The kind of communications I shall be concerned with is, first of all, fast— faster than the means of locomotion at the time, i.e. faster than a horse or a boat in ancient times, faster than a train in the nineteenth century, and faster than an aeroplane in the twentieth and twenty-first centuries. Secondly, it is long-distance communications, meaning that messages are to be delivered at a distance well out of earshot. Thirdly, it is communications from point A to point B. This last distinction has only become significant in recent times. If, say, a Roman Emperor wanted to send a message to a provincial governor he sent a messenger. If the Emperor wanted to send the same message to a dozen governors he sent a dozen messengers. The techniques for sending to one and sending to many were the same. However, modern methods of reaching the many differ considerably from those set up for establishing communications between two persons. In technical jargon the first one is known as broadcasting and the second one as point-to-point communications. I shall keep away from broadcasting (it has too many different facets) and concentrate on the latter, asking the question: how, starting from the earliest evidence, did man manage to send information from point A to point B, far away, without physically delivering the message?

Having limited the subject to be discussed I shall now broaden it. The availability of fast communications has made such an impact upon all aspects of human life that it is impossible to ignore the political and social consequences. I shall discuss them in detail whenever I have a chance. The last and possibly the most important thing I wish to do is not only to describe what happened in the past 4000 years but also to explain the underlying principles as new inventions and new discoveries came along. One might think that the subject of modern communications is far too complicated for the layman to understand. This is true indeed for the past two decades but does not apply to the previous 3880 years. For that period I shall try to describe the operational principles of many of the devices as they entered service.

The Beginnings of Communications

The principal aim of this chapter is to present some evidence of the existence of early communications systems. At the same time, faithful to the dual purpose of the book, the concept of communications will also be discussed starting at the very beginning. Terms like 'binary arithmetic' and 'bit' will be liberally used, and the two digits 0 and 1 will be introduced in the sense used by communications engineers.

In order to emphasize the simplicity of the basic principles it might be worth starting in the world of nursery rhymes. It may be assumed that Jack needs a pail of water but owing to an accident on the previous day he is confined to bed and his head is still wrapped up with vinegar and brown paper. Jill, who lives next door, would be willing to go up the hill on her own and fetch the aforesaid pail of water but she has no idea whether the water is needed.

Jack can call attention to his need in several ways. He can, for example, shout or he can send a brief note. However, Jill's house, particularly when the windows are shut, may be too far away for oral communications, and there may be nobody about to fulfil the role of the messenger. So Jack may decide to send a signal. How to send a signal? Anything that has been previously agreed would do. Using artifices easily available for someone lying in a bed he could, for example, put one of his slippers in the window. According to his agreement with Jill, no slipper could mean 'water is not needed', whereas the presence of a slipper would indicate desire for a pail of water. It is a case of YES or NO; yes, water is needed or no, water is not needed. In the communications engineer's jargon one bit of information needs to be transmitted. YES may be coded by 1, and NO by 0. In the particular communications system set up by Jack and Jill, the presence of a slipper in the window is coded by 1, and the absence of the slipper by 0.

In times less demanding than ours, being able to obtain one bit of information was regarded as quite substantial, particularly in matters of defence. The question most often asked was 'Are hostile forces approaching? Yes or no?' The practical realization of such an early warning system was quite simple. Watchmen were posted at suitable vantage points in the neighbourhood of the city: the watchmen then sent signals whenever they could observe enemy movements. The usual way of sending a signal was by lighting a fire. Lack of fire meant, 'No, no enemy forces are approaching'. The presence of fire meant, 'Yes, enemy forces are approaching'.

Getting the Message: A History of Communications. Second Edition. Laszlo Solymar, Oxford University Press (2021).
© Laszlo Solymar. DOI: 10.1093/oso/9780198863007.003.0002

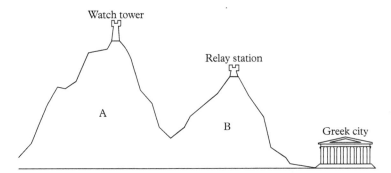

Fig. 2.1 A relay station is needed when those on watch cannot directly communicate with the city.

Next suppose that the fire lit by the men on watch is not directly visible from the city where the information is required. There might be a mountain in between as shown in Fig. 2.1. So what is the solution? Post watchmen on both mountain A and mountain B. Those on mountain A will first see the enemy and light a fire. Those on mountain B will light another fire in turn, and that will be seen in the city. The idea is to have a relay, and there is of course no reason why the relay could not have many more elements—5 or 10 or perhaps 100. In principle, it makes no difference how many elements there are. In practice, there is a higher chance of failure if there are too many of them. At one particular post there might be a flood which makes lighting any fire impossible; at another post the watchmen might be playing dice instead of paying due attention. The various reasons for failures in communications will be discussed at several places in this book.

It would be of interest to know when fire signals were first used. Presumably, as soon as men could confidently ignite a fire, and had acquired some elementary command structure. Documentation is another question. Only a minority of our ancestors were fond of documentaries—and most of those ever written must have perished in the frequently occurring disasters. How far one can go back seems to depend on the diligence of archaeologists and on the ingenuity of those who can decipher odd-looking symbols. It is quite possible that a lot of evidence is still hidden in some unexcavated palaces. As it is, the earliest evidence comes from the middle of the nineteenth century BC.

The city where the evidence comes from is called Mari. Once upon a time it lay on the banks of the Tigris, somewhere halfway down its journey to the Arabian Gulf. It disappeared from history before the close of the century when Hammurabi's forces razed it to the ground. It reappeared in the 1930s thanks to the efforts of a group of French archaeologists. They found an amazing amount of information about the city and about all those with whom the kings of Mari kept up a regular correspondence. The various chambers of the excavated palace yielded over 20,000 clay tablets written in Akkadian. They are particularly informative because in that period the letters were written (using cuneiform writing which had a symbol for each syllable) in the living language. They give accounts of all kinds of activities; for example: register of people obtained from the last census, records of incoming and outgoing goods (including such disparate items as garlic and gold), legal documents on various disputes, commercial transactions, correspondence with foreign rulers, and reports on administrative and political problems, on the state of the roads, on weather conditions, and (luckily for this book) on fire signals.

One might expect that there would be no need to write reports when the signalling system worked smoothly. Letters written to the king would more likely be concerned with difficulties encountered. The following two letters (see Stephanie Dalley, *Mari and Karana, Two Old Babylonian Cities*. Longman, London, 1984) are indeed of that genre:

Yesterday I went out from Mari and spent the night in Zurubban; and the Yamanites all raised torches: from Samanum to Ilum-muluk, from Ilum-muluk as far as Mishlan. All the towns of the Yamanites in the district of Terqa raised their torches in reply. Now, so far I have not managed to find out the reason for those torches, but I shall try to find out the reason and I shall write to my lord the result. But let the guards of Mari be strengthened, and may my lord not go out of the gate.

The second letter has a similar message:

Speak to Yasmah-Addu, thus Habil-kenum. My lord wrote to say that two torch signals were raised; but we never saw two torch signals. In the upper country they neglected the torch signal, and they didn't raise a torch signal. My lord should look into the matter of torch signals, and if there is any cause for worry, an official should be put in charge.

Unfortunately, we do not know whether an official was ever appointed and if so whether his intervention improved the communications network. There is no doubt, however, that fire signals were used, erratically perhaps, in that part of Mesopotamia some 4000 years ago.

The letters found in Mari clearly show how our civilization, which we like to refer to as Western civilization, had one of its roots in those fertile grounds between the Tigris and the Euphrates. Hammurabi's forces soon put an end to Mari's prosperity. The city disappeared from the stage of history by the end of the eighteenth century BC. The fall of Mari did not of course mean that torch signals fell into disuse. Various forms of fire signals were no doubt used for the next twelve centuries, although no detailed descriptions have survived.

Moving westwards towards Asia Minor and Greece, our next stop is at the beginning of the seventh century BC when, quite likely, the works of the great Homer were first written down. It would be reasonable to

expect in those epics a story about a beleaguered city which managed
to summon help by fire signals at some time or another. My classicist
friends tell me that no such story exists in the epic poems per se but they
can instead offer a simile from the *Iliad* on much the same lines.[1] The
subject is Achilles' head adorned by a gleaming, burning flame. The
whole spectacle is arranged by the goddess Athene with the specific aim
of frightening the Trojans. What does that flame look like? According to
Homer:

As when the smoke rises up from a city to reach the sky, from an island in the
distance, where enemies are attacking and the inhabitants run the trial of
hateful Ares all day long, fighting from their city: and then with the setting of
the sun the light from the line of beacons blazes out, and the glare shoots up
high for the neighbouring islanders to see, in the hope that they will come
across in their ships to protect them from disaster—such was the light that
blazed from Achilles' head up into the sky.

Greece is of course the country where all the exciting action takes
place and I fully intend to return to it but it is worth making a little
detour to another source of our civilization, the Old Testament. The
time is early in the sixth century BC, The source is the Book of Jeremiah,
which gives a contemporary account of one of the periodically occurring Middle Eastern crises. Jeremiah is known as a prophet of rather
gloomy disposition, and it must be admitted that his pessimism was fully
justified. Ten of the twelve tribes of the Israelites had already been taken
into captivity never to reappear. The remaining two tribes, Benjamin
and Judah, were threatened by the Babylon of Nebuchadnezzar.
Jeremiah gave a sound warning (6:1):

O ye children of Benjamin, gather yourselves, to flee out of the midst of
Jerusalem, and blow the trumpet in Tekoa, and set up a sign of fire in
Bethhaccerem: for evil appeareth out of the north,[2] and great destruction.

Returning to Greece a century and a half later the next thing to look
at is another product of the Greek entertainment industry, the theatre.
A reference to a chain of fires can be found in one of the popular plays
that drew the crowds in Athens at the time. The date of its performance
is well known: 458 BC. The title of the play is *Agamemnon*, the first one in
the *Oresteian* trilogy, written by the celebrated Aeschylus. As any playwright, he wrote what the audience wanted to hear and to see: a horror
story. The events take place just after the conclusion of the great war at
Troy. Clytemnestra (sister of the fair Helena who caused all the trouble)
seemingly welcomes back her husband Agamemnon but, in fact, she is
busy plotting his demise. She has a double motive: first she still resents
her husband's act ten years earlier of sacrificing their daughter
Iphigenia in order to ensure fair wind for the Greek fleet. In addition, she
is reluctant to tell him of her affair with Aegisthus. There is usually

[1] Book 18 from line 208 onwards.
[2] Babylon was to the east of Jerusalem. The reference is to the north because that was the customary invasion route to Jerusalem. No army liked to march across the desert.

a marked lack of cordiality in the relationship between the lover and the husband, but in the present case this tendency is further reinforced by the fact that Agamemnon's father murdered two of the sons of Aegisthus' father, and served their flesh to the unfortunate father at a banquet. The full story of vengeance exacted and justice perceived is rather complicated. As far as the communications aspects are concerned, the main point is that Aeschylus wanted a new dramatic touch. The play starts with a soliloquy by a watchman. His job for the past twelve months has been to study the sky from the roof of the palace. He hopes to see 'the promised sign, the beacon flare to speak from Troy and utter one word, "Victory!" ' And indeed before he has a chance to finish his soliloquy the shining light does appear. He cries out joyously:

O welcome beacon, kindling night to glorious day,
Welcome! you'll set them dancing in every street in Argos
When they hear your message. Ho there! Hullo! Call Clytemnestra!
The Queen must rise at once like Dawn from her bed, and welcome
The fire with pious words and a shout of victory,
For the town of Ilion's ours—that beacon is clear enough!

A little later, in reply to the questions of the Chorus, Clytemnestra vividly describes how the message came from Troy. She tells them of the chain of fires lit subsequently on the mountain tops of Ida, Lemnos, Athos, Peparethus,[3] Euboea, Messapium, Cithaeron, The Megarid, and Arachneus (see Fig. 2.2). The Chorus is not entirely convinced; they suspect a possibly unreliable divine message, but her information proves to be correct when later in the play a herald arrives and confirms the fall of Troy.

Did the Greeks in those mythical times set up such an elaborate relay between Troy and Argos? Had they done so those mountain peaks would have indeed provided the best choice. Unfortunately, there is no evidence whatsoever for such a link outside Aeschylus' play. It is certainly not in Homer. So did Aeschylus invent them to present an exciting image to his audience? Possibly. By *his* time relays of beacon fires were widely used, as we know from the works of Herodotus and Thucydides, so why not make use of them on the stage?

I have by now amply discussed the transmission of one bit of information and even put it in historical context, so this might be the right place to graduate to two bits. In the example given, Jack's interest was confined to a pail of water. It will now be assumed that Jack might want a loaf of bread as well. So his possible choices are:

(1)   A pail of water but no loaf of bread;

(2)   A loaf of bread but no pail of water;

(3)   Both a pail of water and a loaf of bread; or

(4)   Neither a pail of water nor a loaf of bread.

[3] Peparethus is not in the text that survived but modern scholarship tends to the view (partly from the syntax and partly from the fact that Euboea and Athos are too far from each other) that it was in Aeschylus' original text.

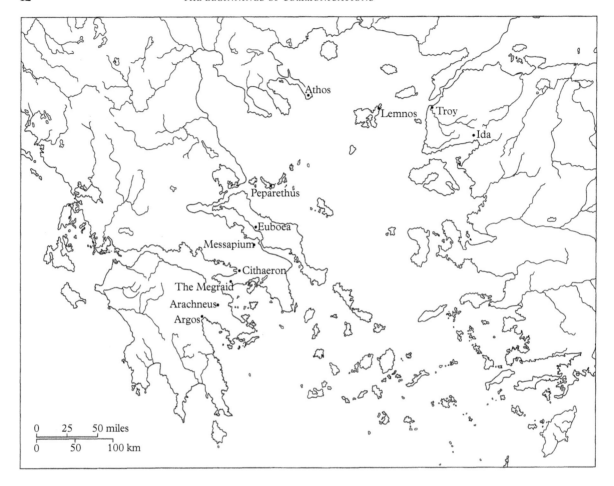

Fig. 2.2 The beacon chain between Troy and Argos.

How could the above options be described in terms of slippers? Now, clearly, two slipper holders are needed. Option (1) could then be coded by a slipper in holder 1, option (2) by a slipper in holder 2, option (3) by slippers in both holders, and option (4) by the complete absence of slippers. Now a particular arrangement of slippers (their presence or absence) would carry two bits of information. It will represent one choice out of *four* possible choices. Using the notation of 1 and 0 for 'slipper present' and 'slipper absent', the options may be presented in the following manner:

Option 1   1,0

Option 2   0,1

Option 3   1,1

Option 4   0,0

The description of the four options is related to 'binary arithmetic', a term that may sound rather intimidating: however, all that needs to be known is that in this example the presence of something is coded with the digit 1 and the absence of the same thing (whatever it is—slippers are not necessary, socks will do equally well) with the digit 0. There are only two possibilities: either something is there or it isn't, so exactly two digits are needed. By the way, it may now be appreciated that 'bit' is not a natural word either: it is a product of the flourishing acronym industry. It stands for 'binary digit'.

My next example will still be a little artificial but bit by bit (if you will excuse the pun) I shall be getting nearer to more realistic coding problems. The assumption is now that there is a language which uses only four letters: A, B, C, and D. According to what has been said already, two bits are needed to describe the four possibilities. Hence the code for the four letters may, for example, be chosen as follows:

A   0,0

B   0,1

C   1,0

D   1,1

The next logical jump is to a language that uses 8 letters from A to H. How might the coding be done now? How many bits are needed? I shall presently show that one more bit, that is the availability of a 1 and a 0, will be sufficient. The code for A to D given above may then be modified by sticking a 0 on to the end. Thus they will take the form

A   0,0,0

B   0,1,0

C   1,0,0

D   1,1,0

Choosing now the third digit as 1 instead of 0 there are clearly four new possibilities which may be used to code the letters E to H as

E   0,0,1

F   0,1,1

G   1,0,1

H   1,1,1

The general rule can now be easily seen. By adding one more bit, the number of possibilities can be doubled. Thus, 4 bits are needed for coding 16 letters and 5 bits for coding 32 letters. If a language does not contain 32 letters (but only 24 as the Greek alphabet or 26 as the English alphabet), then the rest may be made available to code symbols like question marks, exclamation marks, commas, etc.

Now if we put ourselves in the ancient world where for the purpose of communications we have only fire at our disposal, how would we have coded the alphabet? With torches. We would have had 5 designated places which would or would not have displayed a torch.

Did the ancient Greeks think of such a system? Nearly. To find that out it is necessary to leave Aeschylus behind, jump about three centuries and stop at Polybius, one of the best and most prolific historians who ever lived. Unfortunately, out of his 40 books only 5 are extant, but, by good luck, the one in which he wrote extensively about signalling has been preserved. After describing some fairly sophisticated signalling systems (improvements on the simple one-bit message) he comes to one which is capable of sending any message whatsoever. He attributes the invention of this system to Cleoxenus and Democlitus with some further improvements due to himself. The idea is simple and ingenious. The alphabet is divided into groups of five letters as follows:

A	Z	Λ	Π	Φ
B	H	M	P	X
Γ	Θ	N	Σ	Ψ
Δ	I	Ξ	T	Ω
E	K	O	Y	

Since the Greek alphabet has only 24 letters one place remains empty but that is of no consequence. The position of each letter is now determined by its column and its row. For example the letter K is in the fifth row and in the second column. The coding is done by two sets of 5 torches, one set to the left of a mark and the other set to the right. Thus the letter K is coded by 2 torches on the right, and 5 torches on the left. How does this compare with our binary system described earlier? The binary system certainly wins on the number of torches. With five torches we can code any letter out of 32 whereas Polybius needs 10 torches to code one letter out of 25. Does Polybius' system have any advantages over ours? It does. It works much better in the circumstances envisaged when the information is to be relayed by watchmen relying on the power of their naked eyes. In the binary system the relative position of each torch is crucial. As the torches flicker and are swayed by the wind it may not be easy to tell whether it is the second torch or the third one that is missing. Anyway, it is a humbling thought that as many as twenty-two centuries ago communications engineers (Cleoxenus and Democlitus could hardly be qualified by any other description) invented the means of being able to send any message. Polybius' closing remarks on the subject are also quite illuminating:

I was led to say this much in connection with my former assertion, that all the arts had made such progress in our age that most of them were reduced in a manner to exact sciences; and therefore this too is a point in which history properly written is of the highest utility.

Did this communications system ever come into practice? The principles were there but that's about all. To be able to count the number of torches, the watchtowers would have had to be spaced about 1 km from each other. That would have been far too expensive to build and to maintain at the time of Polybius. But with a jump of another 300 years the situation looks much more favourable. In the second century AD Roman emperors were reasonably intelligent and had enormous resources at their disposal. Any communications engineer who would now enter a time machine and resurface in Rome at around that time would certainly pester the sitting emperor to build such a system, and the emperor would very likely give his consent.[4] Surely, such a system would help in the administration of the empire (so the emperor could count on the support of all the administrators), would keep the emperor aware of what was going on in the provinces (any rumour of a revolt?), would give an early warning system against any attack by barbarians, and, even better, would enable the emperor to direct all military operations from his headquarters in Rome. Well, had it happened that way, the Roman Empire might have never collapsed and we would still speak the language of Horace all over Europe. Or, to give another example, a communications system would have given an opportunity to local governors to solicit advice from the central authorities. Pilate, for example, could have sent a telegram to Tiberius asking, 'What shall I do with that turbulent prophet?' And Tiberius might have replied: 'Send him to Rome'. And the prophet might have mellowed with the passage of time and might have been converted in Rome to the worship of Minerva, with incalculable consequences for the subsequent history of the world.

There is no doubt that the Romans were great engineers. Their mechanical engineers produced great war machines, their civil engineers built roads unsurpassed until the eighteenth century, so why were the communications engineers so far behind? The writings of Polybius would have been fairly easily available so it seems quite likely that a number of people had a pretty good idea, at least in principle, of how to build a communications system. That is, however, not enough. There are probably two further conditions to be satisfied. The decision-makers must be aware of both the ideas and the technological possibilities and, secondly, ideas must be tested by *experiments*. One can just imagine an eager young man, who has just finished reading Polybius' *Histories*,

[4] This excursion into antiquity is not my invention. There is a novel by L. Sprague de Camp, *Lest Darkness Fall*, on this very subject. He puts an American in sixth-century Rome where, among other things, he sets up a mechanical telegraph along the Italian peninsula. Under his wise advice the ruling Goths adopt democracy, and thereby mankind is saved from the Dark Ages.

explaining the principles of telegraphy to a decision-maker fairly high up in the hierarchy.

'It won't work', says the decision-maker.

'Why not?' asks the eager young man.

'A message might travel undistorted through two or three relays but mistakes are bound to multiply when the message has to be repeated hundreds of times. One letter is misread here, another letter is misread there, and the message that arrives will be completely garbled. And what about the security aspects? We don't want to tell every Quintus, Marcus, and Alexander of what the Emperor is thinking, do we? And besides, the whole thing would be prohibitively expensive. If I wanted to take such tremendous risks I would rather send a ship to China via the Pillars of Hercules, as one of your friends recently suggested to me. Oh, youth! When I was young...'

The eager young man might have suggested that they set up an experimental line involving only three or four towers and find out the snags, but that would have been against the spirit of the time. You invest a certain amount of technological effort to produce some immediately useful result, like a road for example, or an aqueduct, but you do not waste the effort of so many slaves to produce a white elephant. It was also alien to the spirit of the time to give further thought to possible improvements. If the possibility of distorting the message is a strong argument against building a communications system, wouldn't it then be worthwhile to develop a code that is more resistant to mistakes? If security is a problem, wouldn't it then be worthwhile to develop a code that cannot be easily broken?

Whatever the reason, the fact is that Imperial Rome failed to develop a communications system capable of transmitting any desired message. The need was there, they knew how to build it, they would have been able to afford it, but they just did not do it. They had beacons for early warning but their system was no more sophisticated than that of other empires before them: a couple of watchtowers here, a few watchtowers over there, that was all they ever had. Thanks to the illustrated history of their achievements, left to the world in the form of Trajan's column, it is known what these watchtowers looked like (see Fig. 2.3).

After the collapse of the Roman Empire in the west there was still a chance, a slim chance admittedly, for the Byzantine Empire to develop the telegraph. We know that they did no such thing. In fact, they did not even have chains of beacons apart from a brief interval around the middle of the ninth century. It was a time of frequent Arab raids necessitating an early warning system. Fortunately, it was also the time when Leo the Mathematician (known also as Leo the Philosopher) lived and worked. The system he devised stretched from Loulon, close to the Arab frontier, to Constantinople, 450 miles away, using altogether nine

Fig. 2.3 Roman watch-towers as shown on Trajan's column.

strategically located beacons. The method of signalling relied on clocks situated at the two ends of the chain. The meaning of the fire depended on when it was lit. It meant an Arab raid at hour 1, war at hour 2, arson at hour 3, and some other unspecified event at hour 4. The message arrived in Constantinople about one hour after it was sent. The system worked satisfactorily for a decade or two but then it came to an undignified end some time during the reign of Michael III (842–67). The story goes that a signal indicating an Arab raid came just at the time when the emperor had a winning streak at the horse races. Not wanting to upset the crowd at the next day's race he ordered the dismantling of the system.

From the seventh century onwards the Arabs were of course more than invaders of Byzantium. They founded their own empire. By the middle of the eighth century the Muslim Empire of the Abbasid dynasty stretched from the river Indus in the east to the Pyrenees in the west. Their need for communications was served by a postal system relying on messengers riding horses, mules, or camels, depending on the terrain. The empire was nominally ruled by the caliph sitting in Bagdad but in practice most of the provinces were under local rulers intriguing against one another. Thus, there was some motivation to build a fast communications system for local use. There is only one such communications system on record coming from Arabic sources. The chains of towers, built along the North African coast, apparently covered the enormous distance from Ceuta (opposite Gibraltar) to Alexandria. The chain might have worked for quite a long time, maybe for two centuries. It is known to have been destroyed in 1048 during the revolt of the Arabic West against the Fatimid dynasty (who ruled Egypt at the time). The transmission of fire signals took just one night to run the whole course. The shorter distance from Tripoli to Alexandria was covered in three to four hours.

Europe made no contributions to communications techniques after the victory of the barbarians. It was the time of the Dark Ages, which were not quite so dark for technological development (the saddle, the stirrup, the crank, to name a few, were invented during those centuries and that's also when water mills started to be used extensively). However, the Dark Ages was a complete loss for ideas. Some monasteries might have produced a few eager young men having bright ideas, but no decision-maker would have shown the slightest interest in promoting communications.

Coming to the Renaissance, there must have been renewed interest in telegraphy but apparently only one name has been preserved, that of Girolamo Cardano,[5] an Italian mathematician who lived in the sixteenth century. His proposal is identical to our 5-bit binary system capable of coding 32 letters or symbols. He still relied on torches but in order to distinguish clearly which torches were present and which were absent he proposed building five towers, each of which might or might not display a burning torch. The chances of errors were rather small this way (it should not have been too difficult to observe whether a torch was lit on a particular tower or not) but the cost had to be multiplied by a factor of five, an expensive way of reducing mistakes. The system was, of course, never built and it seems unlikely that any of the rulers of the numerous Italian city states ever gave a thought to it.

It is not my purpose in this chapter to give an exhaustive account of all the beacon systems that ever existed. The list would be too long and not particularly interesting. Fire signals were obviously used by many peoples throughout the ages at times of danger. I would, however, like to give one final example of a chain of fires set up during the reign of Queen Elizabeth I. The aim was to report the movement of the Spanish Armada, on the assumption that their visit was not inspired by peaceful intentions. The means was a chain of beacons set up on the south coast. The event was commemorated by Macaulay, the historian, who also dabbled in poetry:

[5] Mathematicians know him as the first man who found the roots of a general cubic equation.

From Eddystone to Berwick bounds, from Lynn to Milford Bay,
That time of slumber was as bright and busy as the day,
For swift to east and swift to west the ghastly war-flame spread.
High on St. Michael's Mount it shone; it shone on Beachy Head.
Far on the deep the Spaniard saw, along each southern shire,
Cape beyond cape, in endless range, those twinkling points of fire.

CHAPTER THREE

The Mechanical Telegraph

The Beginning of Organized Science

The seventeenth century did not start well for independent thinking. In the year 1600 Giordano Bruno, a believer in the Copernican system, was burnt at the stake. Galileo, another believer, escaped a similar fate only by recanting his views in 1633. However, looking at the century as a whole, it was undoubtably the beginning of modern science. Besides Galileo, Kepler, and Harvey, who entered the century as mature men, it was the century of Descartes, Leibniz, Newton, and Pascal, and of many others like Fermat, the mathematician, Torricelli with his barometer, Boyle and Marriotte of gas laws fame, von Guericke with his electric friction machine and unbreakable hemispheres, Vernier of the Vernier scale, and universal geniuses like Huygens and Hooke.

As for bloodshed, the century was slightly above average. Catholics and Protestants killed each other in great numbers but there was no shortage of Catholics killing Catholics or Protestants killing Protestants. Some people were defenestrated in Prague, an English king lost his head, Spain had her ups and downs (more downs than ups), French cardinals spread intrigue, but science marched on. Inexorably. It even became respectable.

The first organized scientific academy, the Accademia del Cimento, was founded in Florence in 1657 by the two Medici brothers, the Grand Duke Fernando II and Leopold. It flourished for ten years. Tradition says that the end came when Fernando was offered a cardinal's hat and the pope did not think that being a cardinal was compatible with supporting a scientific society. The Académie des Sciences was founded in 1666 in Paris when Colbert, a forward-looking minister of finance, managed to persuade Louis XIV to extend his patronage to the sciences.

The foundations of a scientific society in England were laid on 28 November 1660 when some illustrious scientists met at a lecture by Christopher Wren (the architect of the rebuilt St. Paul's Cathedral). A 'mutuall converse' was held and 'Something was offered about a designe of founding a Colledge for the Promoting of Physico-Mathematicall, Experimentall Learning'.[1] After obtaining the Royal Charter they became known in 1663 as 'The Royal Society of London for Improving Natural Knowledge'. Were the Fellows of the Royal Society concerned in any way with telegraphy? Not particularly: that was not one of the burning questions of the time, but they were interested in all natural and man-made

[1] It may be worth noting that the expected contributions of the founding Fellows were not restricted to matters scientific. On 5 December 1660 they decided to undertake the obligation 'that each of us will allowe one shilling weekely, towards the defraying of occasional! charges'.

Getting the Message: A History of Communications. Second Edition. Laszlo Solymar, Oxford University Press (2021).
© Laszlo Solymar. DOI: 10.1093/oso/9780198863007.003.0003

phenomena about which evidence could be gathered. So they did not dabble in theology but if one of their Fellows thought up a scheme for long distance communications, they were only too happy to consider it.

Telegraphy with a Difference

It fell to Robert Hooke, one of the most inventive men who ever lived, to introduce the subject of 'speedy intelligence' to the Royal Society. There is a brief report just mentioning the subject in 1664, but one can find a lengthy description with figures twenty years later 'showing the way how to communicate one's mind at great distances'. Hooke proposed a symbol for each letter, to be displayed at one site and observed at the next one. He even proposed a mechanism for storing the symbols and for sliding them quickly into the display area. In principle, this was no improvement on Polybius (whom Hooke probably did not read). There is, however, an additional idea that could not have possibly been part of Polybius' scheme, and which did bring the idea nearer to practical realization. The symbols displayed were to be observed from the next site by *telescope.*

So how do telescopes work? They are nothing more than two lenses (or one lens and a curved mirror) so arranged that a faraway object will appear to be much closer. In early embodiments the distance between the lenses could be changed by sliding a tube inside another one. They are a familiar sight in films made about pirates and admirals. Before embarking on a juicy battle, both types are usually seen holding a long adjustable tube to one of their eyes.

When were telescopes invented? Obviously, some time after the invention of lenses which had been used from about the middle of the thirteenth century for correcting eyesight. Progress was, however, rather slow. It was not until 1609 that a Dutch spectacle-maker, Hans Lippershey, cottoned on to the idea of using two lenses a distance apart. Legend has it that he delivered some spectacles to his customers and looking through two of them by chance, he saw the spire of the church in the town of Middelburg brought much closer.

Was Hooke's suggestion of using a telescope a milestone in the history of the telegraph? It was, because it permitted placing the towers much farther from each other, maybe by a factor of ten, yielding a tenfold decrease in the number of stations needed. The chances of building a communications system had suddenly improved. Of course all inventors exaggerate the effectiveness of their invention and Hooke was no exception. He thought that 'with a little practice thereof ... the same character may be seen at Paris, within a minute after it hath been exposed at London'. Hooke did not do any experiments but about a decade after Hooke's report (around 1694) Guillaume Amontons, a member of the

Fig. 3.1 Guillaume Amontons showing his experiments to royal personages in the Luxembourg Garden in Paris.

Académie des Sciences, did actually try such a system in the Luxembourg Gardens in Paris (see Fig. 3.1) in the presence of high royal personages. They were duly impressed but did nothing about it. The time was still too early. Louis XIV was not interested. He could happily govern his country without the need for fast communications.

The Mechanical Telegraph in France

The beginning

Priorities in the history of science and technology are often disputed. Was it X of nationality A, or rather Y of nationality B, who first built such-and-such apparatus? There are no disputes concerning the mechanical telegraph. Everyone agrees that the first communications system able to transmit any information was built by Claude Chappe just about a hundred years after Amontons' experiments. It covered the distance of 210 km between Paris and Lille. The first telegram over that line was sent on 15 August 1794. In principle, there was nothing new in the way it functioned. What was new was that unlike earlier attempts, this one actually succeeded. Why? We must appreciate that the time that elapsed between Polybius' first description of the possibility and Chappe's realization was nearly 2000 years, somewhat longer than the usual germination time for useful contrivances. Why 1794? There was a slight possibility that it could have come earlier but, as discussed before, there were too many obstacles. Could the first communications system have come later? Yes, certainly. How much later? Half a century would be a realistic estimate. Once electricity was discovered and its various manifestations investigated the electric telegraph was bound to come. However, but for Chappe and his brothers and the extraordinary historical circumstances of the time, the mechanical telegraph might have never existed.

There were five brothers, Ignace, Claude, Pierre-François, René, and Abraham who all played a role in the history of telegraphy. The art of transmitting information to a distant observer interested them from early childhood. Claude was destined for the church and might have quietly spent his life between his job of praising God and his hobby of inventing devices. But revolution came and Claude lost his church benefices. Like many others at that time he went to Paris in 1791 to seek his fortune. His brother Ignace was already in Paris as a member of the Legislative Assembly. Claude continued his experiments, and by March 1792 he was in the position of being able to submit a proposal to the Assembly to build a practical communications system. The proposal was referred to the Committee for Public Instruction. The President of the Committee reported favourably on the plans submitted on 1 April 1793.[2] Experiments took place on 12 July 1793 over a distance of 35 km using three stations. A report on the experiments reached the Convention (the successor of the Legislative Assembly) on 26 July, and there and then they adopted the telegraph as a national utility. By 4 August 1793 the Ministry of War was instructed to acquire sites and the system was in working order a year later.

One might make an inspired guess at the factors responsible for success:

(1)   the product was good;

(2)   the inventor was determined;

(3)   the inventor had a brother sitting in the body which decided on the matter; and

(4)   there was a demand for the invention.

Points (1)–(4) were necessary conditions which could have come into play at any time between Polybius and Chappe. The inventor always had to push his invention and of course an invention that does not quite work is of limited interest. The brother, as such, was not necessary but some contacts with the body who control the purse strings have been necessary since time immemorial, and are still not a bad thing nowadays. The demand for information, well, that has always been there.

The real reason for Chappe's success was the timing of his submission. Had he written to Louis XVI a few years earlier, before the rise of the revolutionary tide, the reply would have surely been negative.

[2] We have often referred to the communications systems described as 'telegraphs' but the actual baptism took place only in April 1793 when Miot de Melito, a classical scholar in the Ministry of War, coined the term from the Greek words *tele* (far) and *grafein* (to write). Claude Chappe's original term was *tachygraph* meaning speed-writer.

Regimes which had been running for decades or for centuries are very reluctant to introduce any major change. They would say of the telegraph: 'Oh, yes, it's a nice thing to have, oh, yes, we do believe that it will work but we can't possibly spare the money to set up such a system.' The administrators of the time probably would have believed in the efficacy of the final product more than their counterparts in ancient Rome, but the chances are that they would have been equally reluctant to spend money on it.

In order to put Chappe's proposal in context I shall review here both the political and military situation at the time. There are two things everyone knows about the French Revolution: that the revolutionaries were in favour of Liberty, Equality, and Fraternity, and that the Parisian crowd (known as the sansculottes because of their lack of fashionable wear) stormed the Bastille, a symbol of oppression under the *Ancien Régime*, on 14 July 1789. There was of course a lot more to it. The political situation may be characterized by saying that over the course of the next five years, power moved steadily from moderates to radicals to extremists and then suddenly to philistines.

Needless to say everyone had their own agenda. The moderates (the Feuillants) wanted a constitutional monarchy; the radicals (the Girondins) wanted a republic and a fair amount of social change; the extremists (the Mountain) wanted a completely new start, an eradication of all remnants of aristocratic rule, a new constitution, a new social order, and a new economic policy, and they wanted all these things at once, irrespective of the amount of bloodshed necessary to achieve them.

The wiser aristocrats immediately realized that their future looked rather bleak and took flight. Lots of émigrés congregated outside the French borders and waited for the Revolution to collapse. Louis XVI, a man of indecision, did not know what he wanted. By the time he decided to flee Paris (June 1791) it was too late. He was caught at Varennes, a good 200 km from Paris, and escorted back to Paris.

The kings of Europe looked on with sympathy, and some with foreboding, at the predicament of Louis XVI. The emperor Leopold (it was the last avatar of the Holy Roman Empire, a fairly ineffectual body at the best of times) and the King of Prussia expressed their concern in the form of a declaration at Pillnitz. They declared themselves for monarchy and against disorder. The Parisian sansculottes were displeased. Their basic inclination was just the opposite: against the monarchy and for disorder. There was a war fever cleverly manipulated by the radicals. They were in favour of war because they wanted to put Louis XVI, to use a modern phrase, in a 'no-win' situation. If the war went well their own position would be strengthened, if the war went badly they could blame the King and his contacts with émigrés for the failure. The Feuillant (moderate) Government fell and the radicals came to power.

The war started badly for the French. The Prussians and Austrians crossed the French border. After some initial setbacks the French armies were, however, victorious at Valmy and the attackers had to withdraw. Meanwhile the Parisian sansculottes acquired more and more street power. In June 1792 they stormed the Tuileries where the king was resident, making him a virtual prisoner. The Legislative Assembly was dissolved in September 1792. The Convention was elected in its place and promptly proclaimed the Republic. The radicals suddenly found themselves preaching caution. They passed some radical laws (e.g. the expulsion of all recalcitrant priests, and the abolition of all dues owed to seigneurs, without compensation) but they voted with some reluctance for the execution of the king in January 1793.

The war restarted with new vigour in February–March 1793. France had to face a coalition of England, Holland, Austria, Prussia, Spain, and Sardinia. The French armies had serious military reverses in the first few months. The internal situation shifted towards the extremists. The radicals lost power in June and most of their leaders were executed in the autumn (the revolution started to devour its own children). From the summer of 1793 the extremists ruled. Their power was vested in the Committee of Public Safety which came to dominate the Convention. They governed by terror. The extremists were split in the spring of 1794. Robespierre was in power. The revolution devoured a few more of its children: Hebert was executed in March, Danton in April. A conspiracy against Robespierre succeeded in July. He was executed with his two lieutenants, Saint-Just and Couthon on the 28 July 1794.

How did all these political changes and the continuous state of war affect the telegraph? On the whole it was to the good. Under the *Ancien Régime* the natural thing was to leave things as they were; under the revolution the natural thing was to introduce changes. Obviously, a new thing like the telegraph had a much better chance of support under the revolution. Secondly, the revolution was threatened by external enemies. The revolution had to be saved. The telegraph was intended to help the war effort, so it was a desirable thing to have.

The crucial meeting of the Convention took place on 26 July 1793. Lakanal, a scientist of repute who had witnessed the successful experiments on 12 July, addressed the Convention:

Citizen Legislators,
  The sciences and the arts, and the virtues of heroes characterize the nations who are remembered with glory by posterity. Archimedes, by the happy conceptions of his genius, was more useful to his country than if he had been a warrior meeting death in combat.
  What brilliant destiny do the sciences and the arts not reserve for a republic which, by its immense population and by the genius of its inhabitants, is called to become the nation to instruct Europe.

Two inventions seem to have marked the 18th century; both belong to the French nation: the balloon and the telegraph....

Later he praises experiments:

One does, one fails, one asks questions, one compares, and the positive results come only by experimentation.

Then he praises the population at large:

The inhabitants of this beautiful country are worthy of liberty because they love it and because they respect the National Convention and its laws.

Towards the end he remarks:

I hope you will make good use of the present opportunity to encourage the useful sciences. If you would ever abandon them fanaticism would rule and slavery would cover the Earth. Nothing works so strongly in the interests of tyranny than ignorance.

Summarizing Lakanal's main points: (i) the telegraph was one of the two most important discoveries of the century; (ii) it was a French invention; (iii) it enabled France to teach Europe; (iv) it was an example of the benefits of science; (v) science is good; (vi) the opposite of science is ignorance; and (vii) ignorance favours tyrannies.

Let us now imagine a typical deputy of the Convention. He is proud to be French; he is proud that it was the French nation which shook off oppression and is now leading the way in Europe. He strongly believes that French scientific achievements outstrip those of other countries. He regards it likely that Chappe's telegraph will help to win the war. How will he vote? For the motion, without hesitation. In the unlikely case that some of the deputies have reservations they will have thought twice before voting against the proposal. The radical leaders are already in prison. Who wants to appear to support ignorance and to be on the side of tyranny? The motion is passed unanimously.

Given the historical circumstances, Chappe was bound to receive the commission. The whole enterprise could, of course, still have failed on account of shoddily built apparatus, inferior telescopes, and untrained personnel. But that side was taken good care of by Claude Chappe and his brothers. Everything worked beautifully by August 1794.

What did these telegraphs look like? There are plenty of illustrations, chosen from contemporary engravings and paintings, in the book of Geoffrey Wilson entitled *The Old Telegraphs.* One of these, the St Pierre de Montmartre Church in 1832, with the telegraph erected on the top of the tower is shown in Fig. 3.2a. A schematic drawing, showing the details, may be seen in Fig. 3.2b. The telegraph consisted of a mast about

(a)

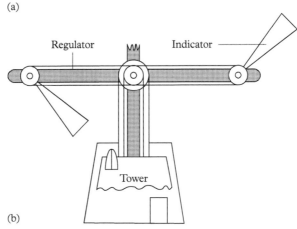

Regulator   Indicator

Tower

(b)

Fig. 3.2 (a) Claude Chappe's mechanical telegraph perched on the tower of the St Pierre de Montmartre church, (b) Schematic representation of Chappe's telegraph.

5 m long upon which a wooden beam, called the regulator, could rotate. At each end of the regulator was an indicator which could also rotate.

There were a large number of possibilities. The regulator could take four different positions, vertical, horizontal, right inclined at 45°, and left inclined at 45°, as shown in Fig. 3.3. Taking the regulator as horizontal,

Fig. 3.3 The four possible positions of the regulator.

(a)   (b)   (c)

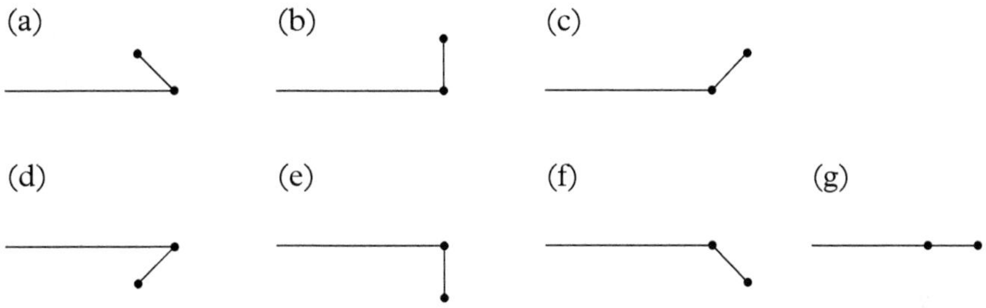

Fig. 3.4 The seven possible positions of the indicator (a further possibility when the indicator is an extension of the regulator was not used).

(d)   (e)   (f)   (g)

the indicator could take seven different positions as may be seen in Fig. 3.4. Notice that the position where the indicator is a mere continuation of the regulator was not used. From a distance it would have been easy to mistake it for the position shown in Fig. 3.4g, in which the indicator is pointing to the left and lying above the regulator. The only difference between them would have been the apparent length of the regulator. In any case they did not need this eighth position as they had plenty of different configurations without it. Considering the 4 possible positions of the regulator and 7 positions of the indicator, there were $4 \times 7 = 28$ configurations with one indicator and $7 \times 28 = 196$ configurations with two indicators.

It would have been perfectly possible to send only letters of the alphabet and numbers from 0 to 9, but the large number of possibilities allowed a more efficient coding devised by Leon Delaunay, a former French consul in Portugal, who knew how to code diplomatic messages. It was essentially a double-code. They had three books, each of them having 92 pages, and each page containing 92 words or more complete expressions, e.g. 'this is the end of the message'. Next, they attached a number to each of 92 different positions of the apparatus. Then a message of 2, 15, 88 meant the 88th word on page 15 of book 2. Coding and decoding occurred only at the terminal stations. The code was not known to the operators.

How many words and expressions could they send by this method? The number of words is clearly $3 \times 92 \times 92 = 25,392$. Was this faster than sending the message letter by letter? Yes, because they could transmit any word by showing 3 subsequent positions of the apparatus, and of course most words contain more than 3 letters.[3]

The speed of signalling for a message was about 1.5 signals per minute. It was rather slow partly because it took time to operate the

[3] There is a question here for those a little more mathematically minded. Why did they use only 92 (actually 98 by including some auxiliary signals) positions out of a total of 196? They could have considerably speeded up the process if they had only one book of 196 pages with 130 words on each page. That would have given them about the same number of words (25,480) without the need to indicate the book. So, they would have saved transmitting one signal, the one that specified the book. Chappe and his collaborators must have thought of such a possibility. Presumably, the errors in decoding the telegrams were higher when they used all 196 configurations. The reduction to 98 was achieved by abandoning the 45° positions of the regulator.

heavy beams and partly because some feedback from the next tower was built into the method of signalling in order to reduce the chances of mistakes.

The modern measure of signalling rate is bits per second. As seen in Chapter 3, 5 bits are needed to send a letter of the alphabet. Taking the average length of a word at 5 letters, each word can be coded with 25 bits. Chappe's telegraph could deliver a word with the aid of 3 signals taking two minutes. Hence, roughly speaking, the signalling speed was 12.5 bits per minute or about 0.2 bits per second. This is in contrast with the figure of about one trillion bits per second that can be routinely transferred today via a single optical fibre.

Further progress

The inauguration of the first telegraph line practically coincided with the fall of Robespierre. There was a backlash in the form of the White Terror and then a fairly quiet period (apart from a minor *coup d'état* in 1797) until 1799, when the country's leadership was entrusted to three consuls with General Bonaparte as the First Consul. In 1804 Bonaparte became emperor as Napoleon I. After many a victory and some mixed fortunes (the Russian campaign of 1812 was particularly painful) he was defeated in 1814 and had to abdicate in favour of Louis XVIII. Napoleon was exiled to the island of Elba where he had a court but not much to do. He returned to France on the first day of March 1815, where he was again enthusiastically received by the crowds, quickly regaining power at home. The rest of Europe, as may be expected, united against him once more. Napoleon remained in power for a hundred days but was finally defeated at Waterloo by Wellington and Blücher. Louis XVIII came back, this time to stay. He was followed in 1824 by Charles X who was swept away by the July Revolution in 1830. Then came Louis-Philippe, the 'citizen King' whose rule was terminated in 1848 by another revolution.

There is no doubt that the establishment of the mechanical telegraph service coincided with one of the most turbulent periods in French history. The interesting thing is that although the turbulence of the age did account for the birth of the service, its subsequent development was practically independent of who was in power. The users, whether they were republicans, administrators in Napoleon's empire, or royalists, all liked to have access to speedy information. The building of telegraph lines went on steadily and irrevocably. The dates for the completion of the various lines are given in Table 3.1. Note that most of them were built in the revolutionary and Napoleonic eras.

As for the Chappe brothers, they remained involved with the administration of the telegraph system with the exception of Claude who, for reasons not entirely known, committed suicide in 1805. The reign of the

Table 3.1 Completion dates of French mechanical telegraph lines.

Paris–Lille	1794
Paris–Strasbourg	1798
Paris–Brest	1798
Lille–Brussels	1803
Paris–Lyons	1807
Lyons–Milan	1809
Brussels–Antwerp	1809
Milan–Venice	1810
Antwerp–Amsterdam	1811
Venice–Rimini–Monte Santa Lucia[4]	1811
Lyons–Marseilles–Toulon	1821
Paris–Bordeaux–Bayonne	1823
Avignon–Bordeaux	1834

Chappes came to an end only after the July Revolution of 1830 when René and Abraham were relieved of their functions. The reason was unlikely to be political. Presumably, someone wanted their jobs. The king, Louis-Philippe, at least showed his appreciation by granting a pension to Abraham Chappe in acknowledgment of his 35 years of service.

A map of the full mechanical telegraph system in France around 1846 is given in Fig. 3.5. By then it was a major enterprise. It had some 5000 km of line with 534 stations.

The last act of the story of the mechanical telegraph was played out under trying conditions (Fig. 3.6) in the Crimean War (1854–6), at a time when British engineers had already laid a 340-mile submarine cable between Varna and the Crimea. The days of the mechanical telegraph were numbered. By the end of 1856 all French mechanical telegraphs stopped waving their arms. The abandoned towers must have provided a dismal sight. Their demise was mourned by Gustave Nadaud:

Que fais-tu, mon vieux telegraphe,
Au sommet de ton vieux clocher,
Sérieux comme un épitaphe,
Immobile comme un rocher.

The mourning continues for eight verses of which the most sentimental is the seventh:

Moi, je suis un pauvre trouvère,
Ami de la douce liqueur:
Des chants joyeux sont dans mon verre;
J'ai des chants d'amour dans le coeur.
Mais à notre epoque inquiête
Qu'importent l'amour et le vin?
Vieux télégraphe, vieux poète,
Vous vous agiteriez en vain!

[4] Note that Brussels, Milan, Antwerp, Amsterdam, Venice, Rimini, and Monte Santa Lucia were under French rule when the telegraphs were built.

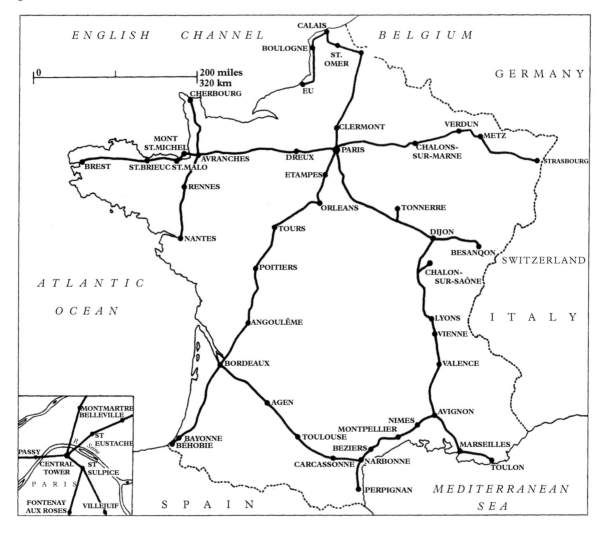

Fig. 3.5 The French mechanical telegraph system *c.*1846.

News carried

The first news on the Paris–Lille line was the recapture of Le Quesnoy on 15 August 1794. The second news of substance was the recapture of Condé-sur-l'Escaut. It was reported to the Convention by Carnot. The Convention, in the best traditions of revolutionary rhetoric, sent back a telegram ordering the change of name of Condé to Nord-Libre, and assuring 'the brave Army of the North' that it 'ceases not to merit well of the country'.

The news was normally of a military nature but not always. The murder of two of the three French delegates to the Congress of Rastatt was reported by telegraph from Strasbourg on 9 December 1797. The news of Napoleon's *coup d'état* against the Directory on 9 November 1799 (18 Brumaire Year VIII according to the Revolutionary Calendar)

Fig. 3.6 The French mechanical telegraph in the Crimean War.

was supposed to be carried by the telegraph, but it did not work that day. As the revolutionary name implies, it was foggy. The news of the birth of the king of Rome, heir to the imperial throne, was carried by all lines on 20 March 1811.

Napoleon used the telegraph widely, both for conducting affairs of state and for directing troop movements. During his various conquests the telegraph lines were often extended well beyond the borders of France to keep Napoleon in touch with the government in Paris.

The rules that only information useful to the authorities should be carried was relaxed for only one item of public interest: the winning numbers of the national lottery. In return, the lottery contributed to the running costs of the telegraph lines.

The telegraph lines were kept quite as busy after the fall of Napoleon, although the nature of the messages changed. On 18 April 1814 the Ministry of Marine Affairs sent a telegram to Boulogne ordering a change of name of the vessel Le Polonais to Le Lys. The reasons may be easily guessed. Louis XVIII was soon expected in the French capital (he did in fact arrive on 2nd May) and any mention of the Bourbon coat-of-arms was likely to find favour with the new king.

The news of Napoleon's landing in France on 1 March 1815 reached Louis XVIII four days later via the Lyons–Paris telegraph line. It was thanks to the telegraph that he knew when to escape from Paris.

On 18 June 1815 the Paris–Lille line carried the wishful message that Napoleon had defeated Wellington and Blücher at Waterloo but a correction was given two days later. On 23 June all lines transmitted the message that both chambers had recognized the infant Napoleon II as successor to his father, followed shortly by the news of the return of the Bourbons. On 31 July the reporting of the winning lottery numbers was resumed. Things returned to normal.

Nationalization

It did occur to Claude Chappe to put the telegraphic service at the disposal of the general public and establish some kind of news service. It was not envisaged as an entirely free flow of information: it would have been censored by the First Consul. The Ministry of War was not interested. It would indeed be difficult to imagine any Ministries of War, then or now, which would be keen to offer their lines of information for private use. However, private lines distinct from the military ones are propositions of an entirely different kind.[5] After the July Revolution of 1830, with a 'citizen King' on the throne, businesses were bound to show some interest in building a communications system. The first to make such a proposal, in June 1831, was Alexandre Ferrier. He thought telegraphs had a great future for the following reasons:

(1)   the advent of railways demands a mode of communications well in excess of their speed;

(2)   stock exchange news should be known far and wide and quickly;

(3)   better communications will lead to higher consumption of industrial goods; and

(4)   the quick transmission of private news is of great value to some individuals.

The arguments were unassailable. In England or in America the official reaction would have been: 'do what you want as long as you don't require any subsidies'. If subsidies were required, then of course the government machine had to be persuaded. This is what Samuel Morse managed to do a little later in the United States. The situation was different in France. French functionaries had always disliked the intrusion of ordinary mortals into 'official business'. Alphonse Foy, the head of the telegraphic administration, was against Ferrier's initiative and he tried to mobilize government support for his views.

Ferrier was wise enough to build his line between Paris and Rouen so as not to duplicate any of the existing official lines. It was ready by the middle of July 1833. Lacking any legal basis for doing so, the government could not ban it but all their local functionaries did their best to make life difficult for the telegraph company. Whether for that reason or for

[5] A notable exception was the US Department of Defense in the early 1960s. The first embodiment of the Internet, the lines between the West Coast and the East Coast, were used by both civilians and the military.

some other, the line stopped working by the end of the year. Undeterred, another entrepreneur built an international line between Brussels and Paris which was opened in October 1836. A Brussels journal, *Le Mercure Belge*, noted soon afterwards that it was now possible to know by night what had happened during the day in Paris.

Foy was determined to put an end to all such enterprises. He persuaded de Gasperin, the minister of the interior, to put in front of Parliament a proposal to set up a state monopoly in telegraphy. The deputies believed that this would not in any way curtail the liberty of the public because the telegraph would never become universal. Compensation was mentioned but the deputies decided that any such talk would only induce dubious enterprises to set up telegraphic lines with the sole purpose of claiming compensation.

The proposal was passed in March 1837 by the Lower House with 212 votes in favour and 37 against, and a month later by the Upper House with 86 votes in favour and 2 against. As already mentioned in Chapter 1, the first paragraph stipulated that

Anyone who transmits any signals without authorization from one point to another one whether with the aid of mechanical telegraphs or by any other means will be subject to imprisonment for a duration of between one month and one year, and will be liable to a fine of from 1000 to 10 000 francs.

Telegraphy on the Other Side of the Channel

The beginning

The French telegraph came into service in August 1794. It did not take long for the news of the telegraph to cross the Channel. No more than a few months later the audience in one of the London music halls was entertained by the following ditty:

If you'll just promise you'll none of you laugh
I'll be after explaining the French Telegraph!
A machine that's endowed with such wonderful pow'r
It writes, reads and sends news 50 miles in an hour.
Then there's watchwords, a spy-glass, an index on hand
And many things more none of us understand,
But which, like the nose on your face, will be clear
When we have as usual improved on them here.

The response of the general public was probably mild amusement. The Navy, the senior service in Britain, was less amused. 'We can't let the French be ahead of us in matters military' was the general approach, and they acted fast. Although a description of Chappe's telegraph was soon available, the Admiralty's choice fell on the so-called shutter

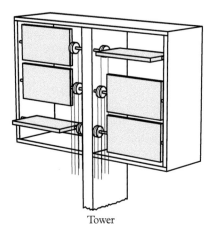

Tower

Fig. 3.7 Schematic representation of the Admiralty's shutter telegraph.

telegraph proposed by Lord George Murray (fourth son of the Duke of Atholl and later Bishop of St Davids and Archdeacon of Man).

In contrast to Chappe's apparatus this was a binary one, each element capable of displaying two positions. The task of the observer at the next post was to ascertain whether any of the 6 shutters were open or closed. A schematic representation is shown in Fig. 3.7.

Preparations were soon in hand. The first two lines were from London to Deal (ready in January 1796) and from London to Portsmouth (ready at the end of 1796). Two further lines to Plymouth (branching off from the Portsmouth line at Beacon Hill) and to Yarmouth were opened in 1806 and 1808, respectively. A detailed map of all four lines may be seen in Fig. 3.8. It is worth mentioning that the delay in the completion of the Plymouth and Yarmouth lines may be directly blamed on the lull in the fighting (there was actually a formal peace treaty concluded at Amiens) in the period from 1801 to 1803. Fortunately for the telegraph, war was restarted in 1803 with renewed vigour.

There is nothing very interesting to say about the telegraphs of the Admiralty. They were useful during the wars and helped to make the blockade of the Continent by the British Navy more efficient. When the Napoleonic wars ended with the Treaty of Paris in 1814, the mood was of general optimism. The victors met at the Congress of Vienna. The occasion was celebrated by sumptuous balls and plenty of dancing, in a manner only the Viennese are capable of.

In contrast to France, where the returning royal administration was as keen as the imperial one to receive news fast, the Admiralty's first thought was to reduce public expenditure. If the war is at an end what's the point of having telegraphs? On 6 July 1814 they gave instructions to dismantle all the telegraphs.

There must have been a few red faces at the Admiralty for shutting down their telegraphs too early. Learning from their mistakes, the

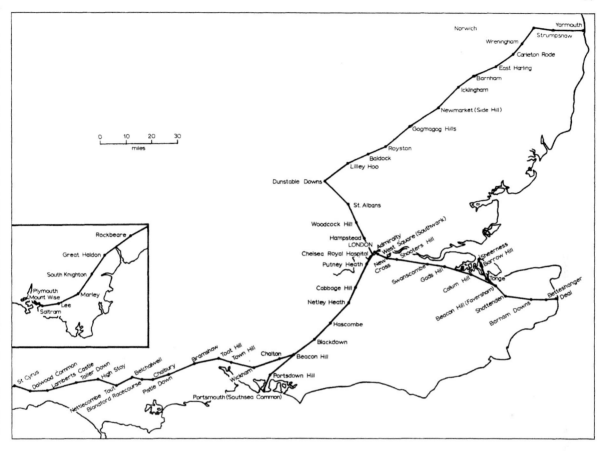

Fig. 3.8 Map of the Admiralty's shutter telegraph system in 1814.

Lordships of the Admiralty were determined that such a thing would never occur again. Let there be peace or war there should always be telegraphs connecting London with the principal ports! And indeed eleven days after the Battle of Waterloo an Act was passed for 'Establishing Signal and Telegraph Stations'. The shutter telegraphs were not resurrected, but a new type of semaphore was installed and was in service until about 1846, when the electric telegraph took over. I shall return to them later in this chapter, when discussing the relative merits of the shutter and the semaphore.

I shall finish this section with the capabilities of the shutter telegraph. Remember, there are 6 shutters and each of them can have two positions: open or closed. This is clearly a binary system with 6 elements, i.e. 6 bits of information may be transmitted corresponding to $2 \times 2 \times 2 \times 2 \times 2 \times 2 = 64$ different combinations. Out of these 26 were taken up by letters, 10 by numerals and 2 fairly obvious ones were related to the beginning and end of a message:

All shutters closed = Message starts
All shutters open = Message ended or not at work

The rest of the combinations stood for often used words like Captain, Admiral, Line of Battleship, Frigate, Convoy, French, Dutch, Russia, Portsmouth, North Sea, Fog, etc.

The commercial telegraph

Liverpool is apparently the place where the first commercial telegraph was set up. It might have started in a rather primitive form in the middle of the eighteenth century when each merchant had a pole on one of the hills. When a ship came in sight a flag was hoisted on the right pole.

The decision to set up proper communications by semaphore was taken in 1825. The Liverpool Dock Trustees were empowered to

establish a speedy Mode of Communication to the Ship-owners and Merchants of Liverpool of the arrival of Ships and Vessels off the Port of Liverpool or the Coast of Wales, by building, erecting and maintaining Signal Houses, Telegraphs or such other Modes of Communication as to them shall seem expedient, between Liverpool and Hoylake, or between Liverpool and the Isle of Anglesea.

The line was set up by a Lieutenant Watson. It was working successfully by October 1827, first reporting the arrival of an American ship which happened to be called Napoleon. The message took 15 minutes to travel.

The first telegraphs erected consisted of 3 pairs of semaphores on a mast of about 15 m height as shown schematically in Fig. 3.9. Each semaphore had three possible positions: horizontal, 45° up, or 45° down. Hence a pair of them had 9 different positions (shown in Fig. 3.10), and the three pairs could exhibit altogether $9 \times 9 \times 9 = 729$ different positions. Again, there was a code book to interpret the various signals.

Watson went on setting up commercial telegraphs from Hull to Spurn Head, from London to the South Foreland and from Southampton to the Isle of Wight, but they were rather conventional without any particular interest.

What is interesting is the contrast between Britain and France. French functionaries, as mentioned several times before, were anxious to keep the monopoly of their telegraph. In Britain the authorities had no objections whatsoever to merchants erecting and maintaining local telegraphic services. In fact, the Admiralty was willing to pay a fee of 10 shillings whenever they needed to send a telegram via Watson's telegraph line from London to the coast.

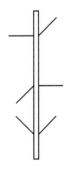

Fig. 3.9 Watson's semaphore.

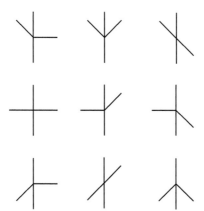

Fig. 3.10 The nine possible positions of each pair of arms.

Mechanical Telegraphs in Other Countries

Once the French military had a communications line at their disposal most European countries were bound to follow suit. In countries under French occupation during Napoleonic times the same kind of telegraphic lines were set up as in France. We have already mentioned such lines in Italy, Belgium, and the Netherlands. As for other countries, they had lines of their own. By the middle of the century all of Europe, from Russia to Portugal, was criss-crossed by lines of the mechanical telegraph. For more details see Appendix 1.

Unauthorized Messages

Was there any chance of using the telegraph for purposes other than those envisaged? It was surely technically possible. It was within the power of any operator to send on a message different from the one received. In Napoleonic times it might thus have been possible for a French party to land in England secretly, raid a telegraph station, and relay some false messages (the code having been obtained previously by spies) to a port, and so send some British ships into a well-prepared ambush. No such French landing party is known to history nor any English raids with the aim of sending false messages through French telegraph lines. What *is* known is a much more orthodox approach to intelligence warfare by Admiral Cochran of the British Navy, who according to a report in 1808

with his single ship the *Imperieuse* kept the whole coast of Languedoc in alarm destroying the numerous semaphoric telegraphs.

It was certainly conceivable to send false messages, although only one single fictional character seems to have ever thought of this. His name is Edmund Dantès, an innocent sailor, who appears in one of the novels by Alexandre Dumas père. He is falsely accused of treason by a bunch of villains and consequently has to spend long years in France's most infamous prison. He escapes, becomes rich, and disguising himself as the Count of Monte Cristo decides to wreak vengeance on his enemies. One of them is Monsieur Danglars, a rich banker, who among other things has a portfolio of Spanish government bonds. The Count bribes a telegraph operator to send on a false report about the arrival in Barcelona of the Carlist pretender to the Spanish throne. The bonds suddenly fall as the news of a likely civil war spreads. The banker has to sell the bonds at great loss and he is ruined.

It seems that only fictional characters ever sent false messages on the French telegraph lines, but there is a true story about the sending of correct messages by clandestine means. The enterprise was conceived and organized by the Blanc brothers, François and Louis, who benefited from it for a couple of years from 1834. They had some accomplices in Paris who inserted some additional signals into the telegraph messages going to Bordeaux. Those who did the official decoding ignored these signals as making no sense, but those in the know deciphered them as giving some crucial information about the Paris Bourse. Unsurprisingly, the Blanc brothers always made a killing on the Bordeaux market—that is, until they were found out and put in prison with all their accomplices. All the accused admitted to sending information on the telegraph and using it for their own gain, but, they argued, that's how the stock exchange worked. Everyone obtained information as well as they could, and used it for buying and selling the right stocks. The accused asked: What exactly was the law they had violated? The prosecution searched in vain for the right paragraph. The accused were acquitted, although they had to pay the legal expenses. One of the brothers (the other one died early) went on to operate on the financial markets of Europe. He died as one of the richest men in France.

The 'Bordeaux affaire', as it became known, did actually play a significant role in persuading the deputies to vote for a state monopoly in communications. Under the new law the Blanc brothers would have been found guilty.

Signal-to-Noise Ratio

A signal is something desired; noise is something undesired. In any practical system there is always noise which will corrupt the signal. Although this chapter has been concerned with signals to be perceived by the eye, it might be good to introduce the concept of signal-to-noise for sound signals. It is easy to give examples. If one conducts a telephone conversation and there is a crackling noise on the line, that is a clear example of noise. The noise, of course, may have nothing to do with the line. A vacuum cleaner used in the same room where the conversation is conducted is another example of undesired noise. How might one define 'noise' for a visual system? If the aim is to see the time on Big Ben, then a vacuum cleaner even at its loudest would not be a source of noise. On the other hand, if in one's line of sight to Big Ben there is a tree which sways in the wind that will certainly represent noise. It will give visual information different from that desired. The best example of noise for visual telegraphy is, of course, fog. Light from the apparatus on the next tower will not reach the observer because it is scattered by the myriads of small particles of which fog is composed.

It may now be worthwhile to go back to the Admiralty's shutter telegraph shown in Fig. 3.7. As mentioned already, the whole system was dismantled after the Peace of Paris in 1814. When in the summer of 1815 the Admiralty decided to set up a new telegraph system, the question arose: what sort? Should it be the old shutter system or a new semaphore[6] with moving arms invented by Sir Home Popham? A trial was held at which the Admiralty's Superintendent of Telegraphs decided in favour of the semaphore. He declared that he could see the semaphore signal with the naked eye whereas he needed a telescope for the shutter.

So why was the semaphore superior? Because it had a better signal-to-noise ratio. If erected correctly, the moving arms of the semaphore would be seen against a clear sky, and would represent a sharp contrast. When one tried to see whether a particular shutter is open or closed, then one was bound to see the wooden framework holding the shutters as well. The light coming from the frame was clearly 'noise'. The contrast was found to be less sharp. Interestingly, the present-day satellite communications systems have the same kind of advantage over noise as the semaphores. A satellite presents a clear contrast against a quiet sky.

Noise may make it difficult to obtain the required information, but that's not the only way noise can harm communications. It can give a signal when none should arrive, a common hazard in all communications systems. Even if we go back to Roman times, receiving a fire or smoke signal might not always have conveyed the right message, as shown in Fig. 3.12. This kind of ambiguity is alleged to have occurred during Elizabethan times as well, when the approach of the Spanish Armada (see Chapter 2) was to be reported. There is a story about a village idiot in Somerset setting fire to a haystack which caused major consternation in Whitehall.

One might say that the possibility of false alarms is a trivial consideration. Yes, of course, it is. Nonetheless, the two possibilities of not receiving a signal when one should, and receiving a signal when one

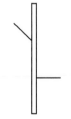

Fig. 3.11 Popham's semaphore adopted by the Admiralty in 1815.

[6]This was an exceedingly simple semaphore with only two moving arms, one above the other as shown in Fig. 3.11. Each arm had seven positions (−135°, −90°, −45°, 0°, 45°, 90°, 135°) yielding altogether 49 different combinations.

Fig. 3.12 Some signals might be ambiguous.

WELL, EITHER THERE'S AN ARMY ON THE WAY, OR THE TOAST IS BURNING!

shouldn't, are two facets of the same problem. Both considerations play a major role in the design of modern communications systems. I shall return to them in due course.

Coding and Redundancy

The Chappe telegraph used codebooks; the English shutter telegraph transmitted individual letters of the alphabet. Which is better? A codebook can, no doubt, speed things up by attaching a long message to a simple signal. A good example of how time can be saved was provided by the German telegraph line. A message saying 'Der König befiehlt' (the King commands) was decoded by the operator as 'Seine Majestät der König haben allergnädigst zu befehlen geruhtet' (His Majesty the King has been graciously pleased to command).

So has the spelling out of words any advantages over the codebooks? The advantage is in the redundancy of the information. Nowadays the concept of redundancy is an important part of the communications engineer's armoury. One might want to make a message more redundant in order to reduce the chances of committing errors, or less redundant in order to speed up transmission.

Let us take as an example the following sentence: 'The telegraph was the first and for many years the most important system of telecommunications'. Now omit all the vowels. It becomes 'Th tlgrph ws th frst nd fr mn yrs th mst imprtnt sstm f tlcmmnctns'. This shows that the vowels are nearly redundant. At first glance one might have difficulties recognizing 'mn' as 'many' but with a little perseverance one probably could get the whole sentence. For the second example it may be assumed that a great many letters in the above sentence are incorrectly received: 'Tha talegbaph was the firsk and fot mahy tears the xost impoytant systev of telecymmonications'. Here, a little more imagination might be needed but the sentence can still be reconstructed. This is because all languages have a certain amount of redundant information which is not absolutely necessary for a correct interpretation of a message. The amount of redundancy in using a codebook is much less. As seen in the discussion of the Chappe telegraph, the signal 2, 15, 88 means the 88th word on page 15 of book 2. If there is an error in coding and the signal sent is 2, 14, 88, then the decoder takes the 88th word on page 14 of book 2 which, needless to say, is quite different. One might of course realize that that particular word is there by error, but it would be more difficult to guess the correct word.

That there is redundancy in the language was recognized very early in the history of telegraphy. For example, the 1797 edition of the *Encyclopaedia Britannica*, a mere two years after the first telegraph was introduced in England, makes the following bold statement in the article headed *Telegraph*:

The grammarians will easily conceive that fifteen signs may amply supply all the letters of the alphabet, since some letters may be omitted not only without detriment but with advantage.

Fig. 3.13 The public house at Putney Heath built on the site of a former telegraph station.

Fig. 3.14 *Domaine de Vieux Telegraph,* a fine red wine preserving the memory of Chappe's telegraph.

Mementoes

No operational telegraph has survived in its original form but some have been carefully restored and can be found in museums at several places in Europe. The only one I have seen is at Chatley Heath in Surrey but I know that the French have one at Haut-Barr in Narbonne and the Germans at Köln-Flittard.

The memory of the mechanical telegraphs still lingers on in geographical names like Telegraph Hill and Telegraph Road. For example, the public house built at Putney Heath on the site of a former telegraph station, is called The Telegraph and is situated in Telegraph Road (see Fig. 3.13).

An unlikely way of guarding the memory of a mechanical contrivance is to name a wine after it. But that's exactly what happened. In the cellars of the better Oxford colleges (and at a few places elsewhere) one can find a fine French wine called Domaine du Vieux Telegraph, a variant of the famous Chateauneuf du Pape. The label does indeed show (Fig. 3.14) an old mechanical telegraph perched on a tower.

Part II
The Beginning of Electrical Communications

4

The Electrical Telegraph

Introduction

The electric telegraph combines in a judicious manner some properties of electricity with the basic principles of telegraphy. Scientists of the purer kind and technologists often disagree concerning the impact of their respective disciplines upon civilization and upon each other. There is, however, no disagreement about electricity. It did not emerge as an answer to an urgent technological requirement. It was science for science's sake. It came about because some people had both lively minds and time to waste.

The history of electricity is as fascinating as the history of telecommunications. It was already known to Thales of Miletos in the sixth century BC that amber when rubbed will attract small particles. However, the first systematic experiments were conducted only much later by William Gilbert, the man of science in the court of Elizabeth I. His book *De Magnete* was published towards the end of his life in 1600, the same time as Giordano Bruno was burnt at the stake. Since the word for amber is *electron* in Greek he called the force *electric* or, to be precise, *vis electrica*. According to the customs of the time he paid respect to both ancient tongues. The new words coined were in Greek whereas the treatise was written in Latin.

In the seventeenth century Otto von Guericke invented the friction machine, which was equivalent to rubbing amber a little more proficiently. There were, of course, many other pioneers of electricity, each one adding a little to the stock of knowledge. The electric phenomena produced were regarded as so spectacular (sparks and shocks) at the time that they were often presented for purposes of entertainment.

To understand the electric telegraph, knowledge of no more than a few elementary properties of electricity is needed. Electricity may be imagined as a fluid which flows in a circuit provided the circuit contains a pump. The flow is called an electric current. The 'pump' is nowadays a battery or the mains. The first battery (called a 'voltaic pile' at the time) was invented by the Italian physicist Alessandro Volta, who presented his device to the Royal Society in 1800.[1]

Electricity can flow in some materials but not in others. Those in which it can flow are called conductors; those in which it cannot flow are called insulators. Examples of conductors are metals like iron or copper; examples of insulators are rubber, porcelain, glass, ceramics,

[1] The term voltage and the unit volt are named after Volta.

Getting the Message: A History of Communications. Second Edition. Laszlo Solymar, Oxford University Press (2021).
© Laszlo Solymar. DOI: 10.1093/oso/9780198863007.003.0004

and plastics. Air is a good insulator, water is not a good one, and sea water is even worse. It follows then that a good insulator is needed if wires are put in the sea, but the insulator can be dispensed with if the wires are in air. The difficulty is, then, that the wires need some mechanical support (that's how telegraph poles were invented) and that support might, particularly when it is wet, lead away electricity from the wires. Hence there is always an insulator (traditionally porcelain) between the supporting pole and the wires. Several wires, properly insulated from each other, may also be twisted together in the form of a cable and buried underground or sunk into the sea.

The properties of electricity mentioned so far all play essential roles in telegraphy but the most important thing that made telegraphy a practical proposition is the relationship between electricity and magnetism. An electric current deflects a magnetic needle put into its vicinity, as discovered by Oersted, a Danish physicist, in 1820. The second important relationship is that an ordinary piece of iron can be turned into a magnet if a current flows in a wire wound around it.

The Principles of Operation

Armed with the knowledge of electricity and magnetism just discussed, we are in a position to design an electrical telegraph. A schematic representation of our invention is shown in Fig. 4.1. There is a battery and a switch at the transmitting end, then a long piece of wire, and a magnetic needle at the receiving end. In Fig. 4.1a the switch is open, there is no electric current and the magnetic needle is in its rest position determined by the Earth's magnetic field. We may now decide to use this device to report the occurrence of a pre-agreed event, e.g. 'Aunt Bertha has arrived'. The quiescent state of the needle would then indicate 'nothing to report'. When there is something to report we simply close the switch at the transmitting end (Fig. 4.1b). Now a current flows, causing the magnetic needle to deflect, indicating thereby that the expected event has occurred. This telegraph is equivalent to a chain of beacon fires. It can transmit one bit of pre-agreed information, e.g. 'Troy has fallen' or 'Aunt Bertha has arrived'.

Let us now be more ambitious and want to send an arbitrary message by spelling out each letter. How can we proceed? Well, if we want to transmit any one of 26 letters it can be done by placing 26 identical circuits side by side. We may then identify the letter A with the first circuit, the letter B with the second circuit, etc. Thus if we want to transmit the word 'bay' for example, then first we switch on circuit 2. We wait a little until the observer at the other end finds out which one of the magnetic needles has been deflected and then we switch circuit 2 off. The letter B has now been transmitted. For the next letter in our chosen word, the letter A, we switch on circuit 1; for the letter Y we switch on circuit 25.

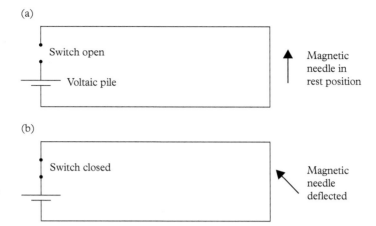

Fig. 4.1 (a) A magnetic needle close to an electric circuit in its rest position. (b) When the switch is closed the magnetic needle is deflected.

Simple, isn't it? But there is a way of doing even better. With our knowledge of binary arithmetic we can immediately propose a solution which would use fewer circuits. Surely for transmitting 26 letters we need no more than 5 circuits. We can then code each letter by various configurations of the five needles. Say, the code for the letter A is 1,0,0,0,0 which in the present context would mean that the first needle is deflected and all the others are stationary. With 5 circuits, remember, we can send 32 different signals (actually, only 31 because the configuration of all needles stationary cannot carry any information).

Can we reduce the number of circuits further? Would a single circuit be sufficient? Yes, because we could transmit our five bits of information as a function of time. How would we code a '1' and a '0'? Well, a possibility would be to press our switch for a long time for a '1' and for a short time for a '0.' So the observer at the other end of the line would have to notice the length of time for which the magnetic needle is deflected between pauses. Then long, short, short, short, short would stand for the letter A. The disadvantage relative to our previous scheme with 5 circuits is that it would now take 5 times as long to send a letter. On the other hand we have tremendous savings in costs. Instead of 5 circuits we need to erect only one circuit.

It needs to be mentioned further that the Earth is a good conductor. The electric current must have a circuit to flow in but the return path could be provided by the Earth.[2] Hence the simplest circuit needs only one wire as shown in Fig. 4.2, with both ends earthed. We have a cylindrical piece of non-magnetic iron around which a wire is wound. The wire is part of an electric circuit provided with a battery and a switch. When the switch is open (Fig. 4.3a) no current flows, the iron is nonmagnetic; hence, it has no effect on the iron plate to which a pen is attached. Both the plate and the spring are in their rest position in Fig. 4.3a. A paper tape which is slowly drawn in the direction of the arrow may also be seen in the figure. When the plate is in its rest

[2] Carl August von Steinheil, following Gauss' idea, tried to use railtracks instead of wires for transmitting telegraphic signals. Owing to the conduction of the Earth the experiments were unsuccessful but it made Steinheil realize that he could use the Earth for the return path of the current.

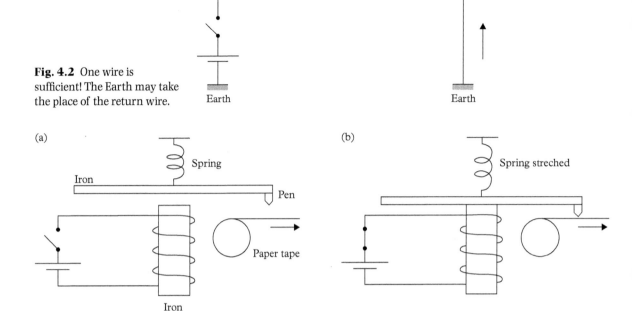

Earth

Fig. 4.2 One wire is sufficient! The Earth may take the place of the return wire.

Fig. 4.3 A piece of cylindrical iron becomes a magnet when current flows in the wires wound around it. (a) Switch open. (b) When the switch is closed the electromagnet attracts the thin slab of iron and the pen descends upon the moving ribbon of paper.

position the pen does not touch the paper tape. When the switch is closed a current flows in the circuit turning the cylindrical iron into a magnet. Consequently, it attracts the plate and the pen is pushed against the moving paper. If now the current is switched off the cylindrical iron is no longer magnetic, it no longer attracts the plate, and the spring can pull back the plate to its rest position. If the electric circuit is switched on for a longer time the pen will draw a dash on the paper. If the switch is on for only a short time a dot will appear. So one could introduce a new code based on dots and dashes, including of course spaces between them.

As the reader will appreciate, we have just invented the telegraph that Morse introduced in 1844. There were, of course, many other possibilities, besides using Morse code, for transmitting information by making and breaking electric circuits. Some of the instruments used were capable of converting the electrical signals directly into letters but that does not affect the basic principles. The relative advantages of the Morse code and of binary coding will be discussed later in the chapter.

The Early History of the Electrical Telegraph

The first proposal to use electricity for sending messages (note that the term telegraph was not as yet coined) was made in an anonymous letter to *Scotts Magazine* in February 1753. The author signed himself as C. M. His proposed telegraph was identical to the first one we designed with a separate circuit for each letter of the alphabet. The only differ-ence was that the electricity was to be produced by a friction machine and the detection by the movement of light balls. Opinions vary as to who C. M. was and why he did not give his full name. One theory says that he was afraid of being branded a magician by his neighbours.

Claude Chappe also experimented with electrical circuits in the early 1790s as contenders for telegraphy, but abandoned them for his mechanical telegraph. Around the same time Francisco Salva in Barcelona proposed a multiple wire scheme. The information was to be detected at the other terminal by the electric shocks experienced by the operators. No wonder that this telegraph never became operational. Von Soemmering exhibited his telegraph in Munich in 1809. He relied for detection on a quite different effect of electric current, namely that it decomposes water. That was not a practical proposition either.

The possibility of using electricity for telegraphy was known beyond the narrow circle of specialists. Some composers of popular verse were also aware of the fact that electricity can deliver a shock. A combination of the two led to the following verse in the *Satirist* (1813):

Electrical telegraphs all must deplore,
Their service would merely be mocking,
Unfit to afford us intelligence more
Than such as would really be shocking.

Next, one should mention Francis Ronalds who lived and worked in Hammersmith, London. He erected two wooden structures 20 yards apart in his garden, upon which he strung 8 miles of insulated wires. With the aid of a friction machine and an elaborate detecting apparatus he was able to transmit information, at least in dry weather (when it was wet he had difficulties with insulation). On 11 July 1816 he wrote to the Admiralty offering his telegraph, stressing its rapidity, accuracy, and certainty. Let us remember that this was the time when the Admiralty had just installed the Popham type of semaphores. The last thing they wanted was another telegraph. In their reply, dated 5 August 1816, they stated: 'Telegraphs of any kind are wholly unnecessary, no other than the one in use will be adopted.' Were their Lordships short-sighted? Not this time. They had a telegraph that worked well and perfectly suited their needs. Why bother about others?

For short-sightedness a much better example would be the view expressed in the 1824 edition of *Encyclopaedia Britannica* under the heading 'Telegraph':

It has been supposed that electricity might be the means of conveying intel-ligence, by passing given numbers of sparks through an insulated wire in given spaces of time. A gentleman of the name Ronalds has written a small treatise on the subject; and several persons on the Continent and in England have made experiments on Galvanic or Voltaic telegraphs, by passing the stream through wires in metal pipes to the two extremities...but there is reason to think that, ingenious as the experiments are, they are not likely ever to become practically useful.

The editor of *Encyclopaedia Britannica* was clearly behind the times because by 1824 the effect of the electric current on a magnetic needle was known. A practical telegraph based on this effect was bound to come, although it was a little slow in arriving.

William Ritchie exhibited such an electromagnetic telegraph in 1830 at the Royal Institution. There was a pair of wires and a magnetic needle for each letter of the alphabet. With so many wires it was obviously too expensive for practical applications.

The first electromagnetic telegraph using a single circuit was probably built by Baron Pavel Shilling, a Russian diplomat, in the early 1830s. Its first public display in the West was in Bonn in 1835. The Russian government became sufficiently interested to ask for a practical demonstration a year later. The experiments were conducted in St Petersburg under quite stringent conditions in and around the building of the Chief Admiralty. Shilling managed to obtain transmission through a cable five and a half kilometres long, part of which was submerged in a canal. It operated without fail for five months. After the success of the experiments the tsar asked Shilling to draw up plans for an undersea telegraphic connection between the naval base at Kronstadt and St Petersburg. The reason for the plan just remaining a plan was the death of Shilling in 1837. However, according to another story, the tsar was never too keen on the project. He was worried that the widespread use of the telegraph might lead to rebellion.

Pioneers with a strictly academic background were Karl Friedrich Gauss and Wilhelm Weber who worked in Gottingen. Gauss was, probably, the greatest mathematician of all time and Weber was a physicist of high repute. They had a brief excursion into telegraphy, as witnessed by one of their papers on magnetic measurements, where they casually mention their telegraph. They set it up between their laboratory and the observatory, drawing the wires over the roofs of the city. For detection they relied on the deflection of a needle to which a tiny mirror was attached. The mirror reflected light so that any slight movement of the needle was observable. Such a device is still in use today, known as a mirror galvanometer.

The first electric telegraph for which companies were prepared to pay money was that of William Fothergill Cooke and Charles Wheatstone. Their transmitting and receiving apparatus was somewhat different from those described already, but there was no difference in principle, so there is not much point going into any detail. However, a few words need to be said about the partnership. Cooke, who up to then had had nothing to do with either electricity or telegraphy, saw a demonstration of Shilling's telegraph at a lecture at the University of Heidelberg in 1836. He immediately realized its commercial significance, built his own version in three weeks, returned to England, and looked for some connections to sell his device. Wheatstone was at the time a professor of experimental philosophy at King's College, London and a Fellow of the Royal Society. He had already conducted a number of telegraphic experiments himself and his model worked reasonably well. The two men first met in February 1837. Cooke proposed a partnership three months later which Wheatstone accepted and they immediately filed a joint patent on 'Improvements in Giving Signals and Sounding Alarms in Distant places by means of electric currents transmitted through Metallic Circuits'.

Why did Wheatstone accept the offer? The reason must have been the same as for any current university professor who has produced something industry might want to buy. He felt at home in the groves of academia but not in the cut-throat world of commerce and industry. He probably regarded it as below his dignity to 'hard sell' his product. Cooke had zeal, ability, and perseverance. He argued, negotiated, shook off rebuffs, tried again and again, and finally managed to land a contract with the Great Western Railway in April 1838, agreeing to set up a telegraph along the railway line from Paddington to West Drayton, a distance of 13 and a half miles. It was actually buried in the ground to protect it from the attentions of the less desirable elements of the public. Operations started about a year later.

Why the railways? Because they needed a means of fast communications for their everyday running. An early problem on the railways was how to tackle a steep ascent. A solution was to attach a rope to the train and haul it up by a stationary winding engine. The distance between the two might have been a mile or two when the rope was attached. It was of great importance to let the machine operator know when the train was ready to move and particularly to send a swift report if one of the carriages got loose. Another example where fast communications are of the utmost significance is single-line working. If the same tracks are used for trains going in both directions, then an accurate knowledge of the position of each train (when they pass which station) is needed.

The partnership between Wheatstone and Cooke was not an easy one. At some stage they even had to go to arbitration to decide who invented what. Cooke went on to commercial success but Wheatstone did not fare badly either. In 1846 he relinquished his royalties in exchange for £30,000, not a negligible sum in those days. The two men also shared the glory. Some time later they were both knighted for their services to telegraphy, Wheatstone beating Cooke by a year.

The effort in America was equally important. As discussed in Appendix 1 there were quite a number of mechanical telegraph lines operating in the US by the 1840s, and with minor exceptions they were all built for the purposes of commerce. The need for new telegraph lines along the railways (or rather railroads as the Americans say) existed in the US just as much as in Europe. So the motivation was there (particularly on account of the great number of railroad lines using a single track), but not as yet a sufficient amount of venture capital. The first electric telegraph line was built along a railway but the money came from the government. The line between Washington and Baltimore was an important one for conducting government business.

The man instrumental in setting up the electric telegraph was Samuel Morse, professor of the literature of art and design at the University College of New York at the time.[3] Shortly after his appointment he began to experiment with the electric telegraph, of which he had first heard a few years before at the end of his European tour. He very soon reached the design whose principles of operation have been discussed and shown in Fig. 4.3. He has been credited with the invention of the Morse code which soon spread all over the world. It was not far from an *optimum* code in the sense that it could transmit a given text in minimum time. He achieved that by the simple method of analysing a certain text and attaching shorter codes to letters occurring more often and longer codes for those coming up less frequently.[4] The single dot, for example, was reserved for *e*, the letter occurring with the greatest frequency.[5]

The US patent for the telegraph was granted in June 1840 but the resources were still lacking for putting it into practical use. Morse asked immediately for a subsidy of $30,000 for the 65-km Washington–Baltimore line. Unfortunately, the postmaster general, fearing competition with the services he provided, did not like the idea. Morse was rebuffed. However, he managed to get support three years later when Congress voted for his proposal by a majority of six votes.

It is quite interesting to compare the two leading powers of Europe, England and France, with the United States. The first one to install the electric telegraph was England, relying entirely on private capital. The year was 1839. Morse opened his line in 1844 with government help. The first French line did not open until 1845, not for lack of technical expertise but because the admirably organized, highly efficient mechanical telegraph network was reluctant to die. In fact, Chappe's signalling system survived for a while. The reason might have been pure sentimentality, although another good reason was to retain the same staff without retraining. Using an ingenious design the detector for the new French electrical telegraph was made to reproduce the movements of the mechanical telegraph, as may be seen in Fig. 4.4.

The French electrical telegraph, just like the mechanical telegraph previously, was set up by the state for the state. As late as 1847 Count Duchatel, the minister of the interior, declared that 'the telegraph

[3] Since Morse was both a painter and an inventor of engineering devices he is very often referred to in the United States as the American Leonardo (except by those who have an acquaintance with the works of the Italian master).

[4] According to another story, Morse deduced the relative frequency of the letters from the lead types of the compositors at the local newspaper.

[5] A very similar code was devised somewhat earlier by Gauss and Weber during their brief flirtation with electrical telegraphy. It was based on the deflection of a magnetic needle to the right (*r*) or left (*l*). The shortest codes were assigned to *e* = *r* and *a* = *l*, followed by *i* = *rr*, *o* = *rl*, etc. Using up to four deflections they were able to code 20 letters and 10 numbers.

Fig. 4.4 An ingenious telegraph receiver which can reproduce the signals of the Chappe telegraph.

should be an instrument of politics and not of commerce'. They relented a little later when the political situation had changed in the aftermath of the 1848 Revolution. The first private telegrams (through the government lines of course; private lines were still not permitted) were sent in March 1851.

Electrical Telegraphy Rises to Fame

The first telegram ever sent (August 1794) informed the Convention in Paris of the recapture of Le Quesnoy. The first telegram sent by the private electric telegraph of Gauss and Weber in Gottingen was 'Michelmann is coming', a much less dramatic message considering that Michelmann was the technician on the job. No great excitement was aroused by the messages carried by the first electric telegraph in actual operation along the Great Western Railway. It was useful for engine drivers, it prevented a number of accidents, helped greatly the efficient running of the railways, but the public was entirely indifferent to it. Very few availed themselves of the opportunity to see the 'wonder of the age', the electric telegraph exhibited at Paddington. Perhaps the entry fee of 1 shilling was a deterrent.

In the absence of war and pestilence it is royalty and crime which arouse the greatest human interest. The telegraph started to be talked about when the birth of Queen Victoria's second son, Alfred Ernest, was announced on 6 August 1844 through a direct telegraph line from Windsor Castle. *The Times* was pleased to attribute the speedy announcement to the 'extraordinary power' of the telegraph. However, even the royal birth was overshadowed by an incident that occurred on New Year's Day in the following year. Initially, the incident looked quite ordinary. A woman was murdered, and as one would have expected it was duly reported in the *Illustrated London News*, a weekly journal whose main aim was to supply the capital with stories of human interest. It was reported briefly because it looked an everyday story of death by poisoning. However, it became clear by the next week that the circumstances in which the suspect was apprehended were far from ordinary. The next week's issue of the journal carried a long article in praise of the electromagnetic telegraph. Among other things, the article gave the full text of the exchange of telegrams between the office at Slough and that at London, Paddington.

THE MESSAGE

A murder has just been committed at Salt Hill, and the suspected murderer was seen to take a first-class ticket for London by the train which left Slough at 7h. 42m.p.m. He is in the garb of a Quaker, with a brown great coat on, which reaches nearly down to his feet; he is in the last compartment of the second first-class carriage.

THE REPLY

The up-train has arrived; and a person answering, in every respect, the description given by the telegraph came out of the compartment mentioned. I pointed the man out to Sergeant Williams. The man got into a New-road omnibus, and Sergeant Williams into the same.

Apparently Sergeant Williams managed to arrest the suspect and produced him at the inquest. A verdict of 'Wilful murder against John Tawell, for poisoning Sarah Hart with prussic acid' was returned. No wonder the case excited great interest. London was in an uproar. The telegraph had arrived. For a while afterwards Londoners referred to the telegraph wires as 'them cords that hung John Tawell'.

In the absence of war, pestilence, crime, and royal news the next candidate for public interest is politics. This has always been of particular interest in the US, perhaps not on a permanent basis, but certainly during the great four-yearly spectacles when the two main parties nominate their presidential candidates. The Democratic Convention in May 1844 was held in Baltimore. By good design Morse's collaborator, Alfred Vail, was at the Convention while Morse himself was in Washington, DC, at Capitol Hill, at the other end of their telegraphic line. As soon as Vail learned who the nominees were (James Polk for president with Silar Wright as his running mate) he sent the news to Morse who got in contact with the prospective vice-presidential candidate. Silar Wright was not pleased. He made it clear that he was not interested in the 'second highest but least important' office in the United States. The reply arrived at the convention a mere half an hour after the decision was taken. The delegates were impressed, the representatives of the press amazed. Not unnaturally, the telegraph rose to fame in the US too.

The Home Secretary Intervenes

In the nineteenth century the English were well behind the French in the number and quality of revolutions. The best they could do was some violence by the Luddites in the second decade of the century, but otherwise everything was quiet until the Chartists appeared on the scene in 1838. This was a working-class movement which presented such outrageous demands as universal manhood suffrage, equal electoral districts, vote by ballot, annually elected Parliaments, payment of members of Parliament, and abolition of the property qualifications for membership. No wonder that such demands were rejected by Parliament although the petitions of 1839 and 1842 contained literally millions of signatures. As a result the movement became more radical. Some violent acts were committed but owing to lack of cohesion among the leaders and to the government's policy of repression, the Chartists did not manage seriously to threaten the established order. But then came the year of 1848. Starting early in the year, the French had their third revolution in 59 years. In March even the Germans, a nation not normally associated with revolutionary fervour, mounted the barricades. The Chartists renewed their offensive. They summoned a convention in April at Kennington Common and prepared another petition. Sir George Gray, the Home Secretary, was not amused. An English revolution no longer seemed impossible.

What could the government do besides the usual measures of sending in troops and mobilizing special constables (mostly from the middle class, but some from the working class were also impressed into service)?[6] They had a new weapon at their disposal: the telegraph. Empowered by the Electric Telegraph Company Act of 1846 the Home Secretary wrote a letter to the chairman of the company, from which I quote below:

[6] It may be worth noting here that Louis Napoleon Bonaparte, the future emperor of the French, had been living in London when the first news of the revolution in Paris reached him. He immediately sailed to France and offered his services to the provisional government. The offer was flatly refused. Returning swiftly to England he reappraised the situation and came to the conclusion that, after all, revolutions might not be such good things. He joined the Special Constabulary in the London area.

I, Sir George Gray ... do hereby authorize and require you ... to take possession of all the Telegraphs and Telegraphic Apparatus of the various stations ... and in so doing to obey only such orders as may from time to time be given to you by me or by my under Secretary ... Given under my hand at Whitehall this 10th day of April, 1848.

The aim was clear: obstruct the lines of communications between Chartists in London and in the provinces, and at the same time ensure and expand communications between agents of the government.

Kingston, Clapham, Harrow, Watford, Wolverhampton, and Hampton, all areas the government could count on, were provided with extra telegraph stations. As it happened the Chartist demonstration fizzled out. Did Britain avoid a revolution on account of the telegraph? It seems unlikely that many historians would seriously entertain the idea but there is no doubt that the telegraph considerably helped those in power.

Progress in England from 1839 to 1868

The first telegraph line, as mentioned earlier, was opened in 1839; the second one along the Blackwall Railway in 1840. The Electric Telegraph Company was registered in 1845 by Cooke and his solicitor, Robert Wilson. The monopoly of the Electric Telegraph Company was challenged by a new company in 1849 and yet another company in 1851. These latter two companies amalgamated in 1857 to form the British & Irish Magnetic Telegraph Company, which became known as the 'Magnetic'. In 1859 the London District Telegraph Company was formed with the explicit aim of wiring up London over the roofs. The last of the major companies, the United Kingdom Telegraph Company (UKTC), went into business in 1860 with plans to erect lines along public highways.

The telegrams travelled from one city to another provided the two were directly connected by wires. But a telegram from Birmingham to (say) Bristol would have been sent in the early days via London. The message would have been decoded in London and then sent from London to Bristol. That took time. Delivery was by special messengers but that took time too. Messages sent from one end of the country to the other might still have taken five or six hours to arrive. If two organizations wanted quickly to contact each other the simplest solution was to lay a line between them. The Universal Private Telegraph Company was formed in 1861 with exactly that aim. They had a number of distinguished clients, the royal family being among them.

To see how the telegraph became universal it may be best to follow the fortunes of the 'Electric', the biggest one among the companies. Their financial position varied a lot. They did not pay any dividend until 1849 and even then it was only 2 per cent. From then on things looked up. Gross receipts and net profits for the period 1851 to 1867 are plotted in Fig. 4.5. It may be seen that they both increased—inexorably. By 1867 net profits represented 42 per cent of gross receipts. Dividends varied between 6 and 8 per cent for the years from 1851 to 1863, reached 10 per cent afterwards, which was the maximum allowed by an Act of Parliament. The total length of wires erected also increased significantly as shown in Fig. 4.6a. However, the greatest increase achieved was in the messages transmitted as may be seen in Fig. 4.6b.

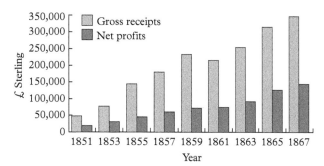

Fig. 4.5 Gross receipts and net profit of the Electric Telegraph Company in the period 1851 to 1867.

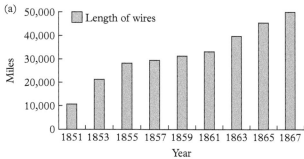

Fig. 4.6 (a) The length of wires laid and (b) the number of messages transmitted by the Electric Telegraph Company in the period 1851 to 1867.

One might conclude from the spectacular rates of growth that the shares of the company sold like hot cakes. They did not. Investors regarded telegraphy as a risky business. The companies had hardly any assets: the lines and the instruments and that was about the lot. Moreover, the lines represented not only assets but serious liabilities too. They had to be maintained. Snow, gales, and flood took their toll. In 1856, after seven years of paying dividends at or over 6 per cent, the price of the shares, issued at £100, was only £80. Even in 1865 when the dividend rose to 10 per cent, the shares were still selling only around £125, offering an 8 per cent yield.

As may be expected the charges steadily declined with time. For 20 words sent to a distance of 100 miles the 'Electric' originally charged 4s. 2d., which was reduced to 2s. by 1855 (see the resulting steep increase in the number of messages in that year in Fig. 4.6). The overall charges in 1858, both inland and to the Continent, may be seen in Fig. 4.7. Amsterdam and Rotterdam could be reached for a mere 6s. The more adventurous had to pay more. St Petersburg cost £1. 11s. 6d. and Constantinople, the most expensive one, as much as £1. 13s. 6d.

The 'Electric' had the greatest market share. Next came the 'Magnetic', then the UKTC, and finally the District, which operated in London so they could carry lots of messages on relatively few wires. The distribution of the lengths of wires and of the messages between these four companies in 1868 are shown in the pie charts of Figs. 4.8a and 4.8b.

THE ELECTRIC AND INTERNATIONAL
TELEGRAPH COMPANY.

INCORPORATED 1846.

The Charges for Messages not exceeding Twenty Words in Great Britain and Ireland :—

Within a Circuit of 50 Miles	....	1s. 6d.
Do. do. 100 do.	....	2s. 0d.
Do. do. 150 do.	....	3s. 0d.
Beyond a Circuit of 150 do.	....	4s. 0d.
To or from Dublin	....	5s. 0d.

No Charge is made for the Names and Addresses of either Sender or Receiver, or for Delivery within half a mile of the Company's Offices. The Company have

UPWARDS OF 360 STATIONS IN FULL OPERATION,
The whole of which are in
Direct Communication with the Continent,
Viâ the Company's
LINE TO THE HAGUE AND AMSTERDAM;
By which, under recent arrangements with the Continental Governments,
GREAT REDUCTIONS
Have been made in the charges, as shewn in the following list of
CHARGES TO

	£ s. d.		£ s. d.		£ s. d.
Amsterdam	0 6 0	Copenhagen	0 12 0	Paris	0 11 0
Antwerp	0 7 6	Genoa	0 15 0	Riga	1 5 0
Berlin	0 11 0	Hamburg	0 10 0	Rotterdam	0 6 0
Bremen	0 8 6	Konigsberg	0 13 6	St. Petersburg	1 11 6
Brussels	0 7 6	Malta	1 11 0	Stockholm	0 18 0
Christiania	0 18 0	Memel	0 13 6	Trieste	0 12 0
Constantinople	1 13 6	Odessa	1 11 0	Vienna	0 12 0

For information as to number of words allowed, charges to other Stations, &c., &c., apply at any of the Company's Offices.

PRINCIPAL STATIONS IN GREAT BRITAIN.

Aberdeen	Falmouth	London	Sunderland
Birmingham	Glasgow	Lowestoft	Swansea
Bradford	Gloucester	Manchester	Truro
Bristol	Greenock	Newcastle-on-Tyne	Wakefield
Cambridge	Halifax	Norwich	Warrington
Cardiff	Haverfordwest	Oxford	Whitby
Carlisle	Holyhead	Perth	Wolverhampton
Darlington	Huddersfield	Plymouth	Windsor
Dublin	Hull	Preston	Wigan
Edinburgh	Leeds	Sheffield	Yarmouth
Exeter	Liverpool	Southampton	York.

Lothbury, London, June, 1858.   J. S. FOURDRINIER, Secretary

1-10-Lo.-11.

Fig. 4.7 The charges of the Electric and International Telegraph Company as for June 1858.

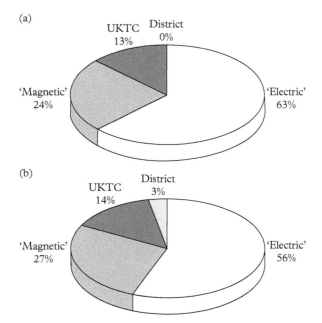

(a)

UKTC
13%

District
0%

'Magnetic'
24%

'Electric'
63%

(b)

UKTC
14%

District
3%

'Magnetic'
27%

'Electric'
56%

Fig. 4.8 The distribution among the telegraph companies of (a) the lengths of wires and (b) the number of messages transmitted in the year of 1868.

News Agencies

Newspapers needed news items to print. Commercial organizations needed information to make money. The idea that both needs could be supplied by an agency took root in the first half of the nineteenth century and burst into blossom with the advent of the telegraph. Bureau Havas was founded in Paris in 1832, Wolff's Bureau in Berlin in 1849, and Reuter's in London in 1851.

Paul Reuter was a Prussian immigrant with some previous experience in newsgathering in Paris and in Aachen.[7] He started up in London by providing opening and closing prices at the London and Paris Stock Exchanges for his business clients. Political news soon followed. A substantial step forward came in 1858 when he managed to secure subscriptions first from a venerable newspaper, the *Morning Advertiser* (founded in 1794), and closely afterwards by the newcomers *The Daily Telegraph*, the *Daily News*, the *Morning Star*, and the *Evening Star*. *The Times* held out a little longer. Its editor regarded it as below dignity to use anything but his own correspondents but finally he succumbed too. He agreed to pay two shillings and sixpence for 20 words if Reuter's name was acknowledged as the source of the report, and five shillings if it was not.

Reuter's greatest coup came on the 7 February 1859. The occasion was Napoleon III's address to the French Legislative Chamber. A hardline speech was expected preparing France and the rest of the world for the announcement of hostilities against Austria. Reuter daringly asked for the privilege of receiving a copy of the speech before it was actually delivered. The emperor, not averse to a bit of publicity, gave his permission

[7] At one time Reuter provided a carrier pigeon service between the end of the Prussian telegraph line in Aachen and the French telegraph line in Brussels.

stipulating only that the envelope should not be opened before the speech began. The transmission of the speech was assured by buying a full hour on the Paris–London telegraph line. Translation was done speedily in Reuter's office in London. Special editions of the London papers were on the streets within an hour of the emperor finishing his speech. Reuter's name became well known.

When war broke out between Austria and the French–Sardinian alliance, Reuter made the news by having correspondents with all three armies. On one occasion he gave to the London newspapers telegrams concerning the same battle from the three different sources.

His name became well known but that did not necessarily mean that it could be properly pronounced. An authoritative view came from the pages of the *St James's Gazette*.

I sing of one no Pow'r has trounced,
    Whose place in every strife is neuter,
Whose name is sometimes mispronounced
    As Reuter.

His web around the globe is spun,
    He is, indeed, the world's exploiter:
'Neath ocean, e'en, the whispers run
    Of Reuter.

Who half so well resolves a doubt?
    When tact is needed, who adroiter?
I trow Earth could not spin without
    Its Reuter.

While Reuter prospered in England, so did his competitors on the Continent. There was rivalry between them but it never degenerated into open warfare. Their conduct was always guided by common sense and ad hoc agreements. For a while Reuter made some inroad in the Hanseatic cities but could not hold out for long. In January 1870 a treaty dividing up the world was signed between the agencies.[8] To Wolff went Scandinavia and Central and Eastern Europe; to Havas the French Empire and the Latin countries of Europe and South America; and to Reuter the British Empire and the Far East. The agreement worked for nearly half a century.

Nationalization

The telegraph was not long in existence, a mere 16 years, when a review article appeared in the *Quarterly Review* published in London. It was a time when science was still regarded as experimental philosophy and the two cultures had not as yet separated. The major part of the article was devoted to the technical aspects of the telegraph which, presumably,

[8] The only previous attempt to divide the world was much less successful. It took place in 1492 when Alexander VI, the Borgia pope, was asked to arbitrate. He drew a neat line beyond the Azores on the map. All new discoveries to the west went to Spain, all those to the east to Portugal. He was apparently oblivious of the existence of France and England.

were of interest at the time to men of education. The tone was fairly admiring. Electricity, which made the telegraph possible, was compared to 'a spirit like Ariel to carry our thoughts to the uttermost ends of the earth', and was described as a 'workman more delicate of hand than the Florentine Cellini, and more resistless in force than the Titans of old!'

Economic matters were also considered, in particular what would be called today the elasticity of demand for telegraphic services. The authors looked at the increase in the number of messages sent as the price dropped at various times in the past, and they concluded that there was ample room for further reduction in price. In the same spirit, they advocated the introduction of a tariff independent of distance. They believed that competition between the companies was not in the public interest because it would only be competition for the lucrative markets in the big cities leading to duplication of effort.

Next, they opened Pandora's other box which contained only two pieces of paper with the words 'Nationalization' and 'Privatization' written on them. The first one escaped, the second one was firmly shut in for another century and a quarter. They asked: 'Would it not be wise for the legislature to consider the question of telegraphy in England before it is too late?' And they asked further: 'Is not telegraphic communication as much a function of Government as the conveyance of letters?'

The questions asked above represented the first salvo in the war between the advocates and opponents of nationalization, which was to last 16 years. What was the line-up? Having got used to the political culture of the twentieth century one might think that the arguments followed party lines, that the Liberals, the party of progress, were for it, and the Conservatives, defenders of vested interests, were against it. In fact, the approach of both parties was purely pragmatic. Both would give their blessings to nationalization provided the public was in favour. Was the world of commerce against nationalization? Not at all. Most chambers of commerce were for it, so were the newspapers, the Post Office, and public opinion in general. Who then were against it? Only the telegraphic and railway interests.

It is easy to explain the stand of the general public, of the chambers of commerce and of the newspapers. They wanted a cheaper service and they hoped the government would provide it. The Post Office was for it because it hoped for increased profit. The opposition of the telegraph companies was of course expected. The main reasons for the railway interests to oppose the plan were threefold: (i) many of the directors of the railway companies had business interests in the telegraph companies; (ii) the railways were beneficiaries of the status quo because they received rent from the telegraph companies for allowing them to set up wires along the railway lines; and (iii) they did not cherish the idea that a public utility could simply be taken away from the shareholders by the government. After all, the railways were also public utilities, and might have possibly come under threat at a later time.

The opponents of nationalization could not, of course, use arguments based solely on self-interest. They tried to win support for their cause by claiming that in the long run the whole thing was against the public interest. They claimed that any lowering of the tariff would be paid for by higher taxation, that there would no longer be any inducement to improve the service, and that it would 'make the press dependent on the whim or favour, or perhaps prejudice, of a Government official'. The arguments were quite ingenious considering that there was no evidence that any of these things might happen. The only nationalized industry at the time was the Post Office, which worked to the satisfaction of all, and far from being a burden on the public purse, it produced a handsome revenue.

There is no doubt that the opposition fought bravely but after a while they realized that they were going to lose. They decided to sue for peace, to surrender, provided the conditions were right. By right conditions they meant bags and bags of money.

That raises the question: how much is a company worth? A tentative answer might be the market value, as determined by the price of shares before the bid is made. In practice, of course, it occurs very rarely that a company can be acquired at its market value. Once the bid has been announced the shareholders will hold out for better terms. If the bidder is the government, shareholders will hold out for even better terms.

The share price of the 'Electric' has already been quoted as £125 in 1865. The increasing certainty of nationalization drove the share price up.[9] It stood at £132 in January 1867, at £145 in November 1867, at £153 in January 1868, and at £255 in July 1869. The shares of the 'Magnetic' doubled during the same period, whereas those of the UKTC quadrupled. Reuters, who also had considerable telegraphic interests, fared best. Their shares rose from £16 to £70.

In 1867 the Post Office's estimate for nationalization stood at 2.4 million pounds, rising to 3.1 million in February 1868, to 4 million in April, and to 6 million in July when the bill passed through its final stages. The total cost eventually reached £10,947,173. Out of this £7.8 million was paid to the telegraph and railway companies, £2.1 million was spent on extensions, and £190,000 was paid as compensation for interests disturbed.

May one conclude that the Post Office was taken to the cleaners by the telegraphic interests? The answer is yes, that would be a fair description of what happened. In France, comparable lengths of wires were built by the state system for a sum of less than a million pounds. They did not allow any private lines. What is better? To muddle through as is usually done in Britain or to make a brave decision and stick to it as is usually done in France? Concerning the inland telegraphic service the

[9]The Postmaster General gave official notice on 15 November 1867 in the *London Gazette* that he intended to purchase the telegraph companies.

advantage was with France, no doubt. They started late but soon caught up and they managed to do the whole thing at a fraction of the cost incurred in Britain. The telephone service fared equally badly in both countries but in the field of submarine cables British industry reigned supreme. So what is better? Heaven only knows, though it might be worth mentioning that a 'dirigiste' approach by the French did succeed in the 1970s. The story will be told in Chapter 11 which is concerned with digitalization.

Progress in Britain from 1870 to 1911

As soon as the companies saw the writing on the wall, any further development stopped. The years of 1867 to 1870 were those of stagnation except that the Post Office, in anticipation of the takeover (it took place officially on 28 January 1870), started to build extensions. As a result the Post Office was able to begin with about 1,000 postal telegraph offices and about 1,800 offices at railway stations. In the first year 91 per cent of the messages were transmitted from post offices, showing that the transfer was for the benefit of the public. The biggest advantage to the public was, of course, the new flat rate. The charges were fixed at 1s. per 20 words independent of distance, the names and addresses being free.

During the first year the number of telegraphic offices increased by about 50 per cent in the major cities. In London the increase was from 95 to 334 offices. By 1872 the system comprised more than 5,000 offices including the railways, and the length of wires increased to 83,000 miles. The increase in the number of messages sent was even more spectacular as may be seen in Fig. 4.9. The average yearly increase is a remarkable 15 per cent.

Profits were expected to decline initially but to recover after a time. The reason for optimism came from the supposed analogy with the penny post. When that uniform rate was introduced in 1839 the profits of the Post Office first plummeted but then slowly recovered reaching the 1839 level 24 years later.

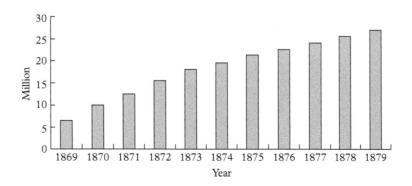

Fig. 4.9 The number of messages transmitted in the period 1869 to 1879.

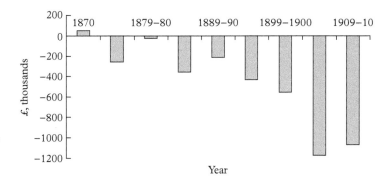

Fig. 4.10 The net revenue of the telegraph service after nationalization up to 1910.

In the first year after nationalization net profits were £342,000, just about enough to cover the interest payment on the £10 million capital stock. There was, in fact, a net revenue of £47,000. Unfortunately, in the years following profits did not live up to expectations: they showed a downward trend. For the period 1870 to 1910 the net revenue (net profits less interest payments) is plotted in Fig. 4.10.

The major pressures on the telegraph department had become clear by this time. The public, aided and abetted by the press, demanded better service at a lower price. The employees demanded higher wages. In contrast to private industry the department could never claim that the new demands would bankrupt them. They lowered their charges (the uniform rate was reduced to 6*d.* in 1886), they paid higher wages, and the debts kept on accumulating.

Labour relations were not entirely harmonious, but good on the whole. There was only one partial strike in October 1871. In a wage settlement in August 1872 the average annual wage of a male telegraphist in London increased to £68 and that of a female to £46. There were further rises in wages in 1885, 1891, 1897, 1905, and 1908, adding to the wages bill the amounts of £129,000, £87,000, £82,000, £95,000, and £210,000, respectively.

Women's Employment

Much has been written about the destitution of factory workers in Britain in the first half of the nineteenth century. Working-class women could hardly live on the wages earned. For women with education the only outlet for their talents was to take on a job as a governess. The electric telegraph changed all that by offering relatively well-paid white-collar jobs to women. After the first woman employee in 1855, the number of women at the Electric's central office grew to about 100 by 1859. The women worked in comfortable, clean surroundings (Fig. 4.11). For the company it was worthwhile because the women were reliable, tended to change jobs less frequently, were less likely to organize themselves, and, above all, were paid less than their male counterparts for the same job.

All telegraph companies ran a training programme for young ladies. Those who could not reach a speed of 8 words per minute were dismissed. Those who could were employed at 8s. per week. The wages could rise to 15s. for a more experienced operator and might have reached 30s. for the few who could really signal fast (approaching 30 words per minute). Their hours were not too long by the standards of the time: from 9 a.m.

(a)

(b)

Fig. 4.11 (a) Women telegraph operators at the Central Telegraph Office. (b) Telegraphic operators at the Central Telegraph Office in 1892.

to 7 p.m., six days a week. Food was consumed on the premises, usually provided by a cook from the raw material the women provided themselves.

When the Post Office took over the private telegraph companies they kept the female staff. In fact, in his 1872 report the Postmaster General emphasized that he had 'given approval to the measures that had been proposed for increasing the employment of women in the Post Office'. Elsewhere in the report he comments:

From the first day of the transfer, the Department entered on the experiment of employing a mixed staff of male and female officers; and there has been no reason to regret the experiment. On the contrary it has afforded much ground for believing that where large numbers of persons are employed, with full work and fair supervision, the admixture of the sexes involves no risk, but is highly beneficial. It raises the tone of the male staff by confining them during many hours of the day to a decency of conversation and demeanour which is not always to be found where men alone are employed. Further, it is a matter of experience that the male clerks are more willing to help the female clerks with their work than to help each other; and that on many occasions pressure of business is met and difficulty overcome through this willingness and cordial co-operation.

The London Central Telegraph Office looked indeed a happy place with a mixed staff of men and women as shown in Fig. 4.11b. Satisfied with the experiment the Post Office not only continued to employ women but employed them at an ever-increasing rate even before the advent of the telephone exchanges. In the period 1884 to 1900 the male staff related to telegraphic duties doubled in both the metropolitan and provincial areas. During the same period the female staff grew by a factor of two and a half in the metropolitan areas and quadrupled in the provinces. The really great increase came later when the Post Office took over the telephone companies. That will be described in Chapter 5.

Submarine Telegraphy: From the Channel to the Atlantic

The possibility of sending electric signals across rivers by insulating the wires was already known towards the end of the eighteenth century. By 1850 large number of successful experiments were done by, among others, Francisco de Salva in Spain, Charles Wheatstone in England, Baron Pavel Shilling in Russia, and William Brooke O'Shaugnessy in India. However, the first solution of the insulation problem suitable for mass production was due to Werner von Siemens of the Royal Prussian artillery. He used gutta-percha (a material similar to rubber which had just then become available) as the insulator and he also invented a method for coating the wires.

So everything was technically ready for conquering the seas. Motivation and capital were still lacking. For establishing a telegraphic communications system within any European country the lines might have had to cross the odd river but the seas could be left in peace. Even when it came to extending communications beyond the borders, all European countries could easily do so—except Britain. The initiative to lay cables under the sea clearly had to come from Britain. Where was the capital to come from? It was a major project so it was logical to expect the British government to step in with a subsidy. After all, it was in the government's interest to be able to communicate beyond the country's borders. The brothers Jacob Brett and John Watkins Brett certainly thought so. In 1845 they wrote to the prime minister, Sir Robert Peel:

We beg the honour to submit a plan for general communication by means of oceanic and subterranean inland electric telegraphs for which patents have been secured by the undersigned and for their construction of cheap and efficient plans.

The Government was not interested. Undeterred, the Brett brothers raised the funds from private capital. Their first attempt in 1850 to lay a cable across the Channel was successful and led to an immediate exchange of greetings between Queen Victoria and Louis Napoleon, who was by then the president of the French Republic. The subsequent failure within a few hours was blamed on a French fishing vessel. According to one story the fisherman caught the cable in his net, cut off a piece and showed it to his friends as a new kind of seaweed with a golden centre. Attempts to recover the cable were unsuccessful. The cable laid in the following year worked for 37 years without any fishermen ever catching it.

Britain's splendid isolation was over. Was it regarded as a good thing to be connected to France? That was certainly the opinion of *Punch* (Fig. 4.12). They thought the cable would promote 'Peace and Good-Will' as the cartoon printed at the time shows. They happened to be right. Never since have France and England taken up arms against each other.[10]

Other parts of the Continent were also shortly to be connected. By 1857 there were direct communication links with Holland, Germany, Austria, and Russia. A cable to India was high on the agenda of private companies. The alternatives were a landline through the Ottoman Empire or a submarine cable from Suez. The British government favoured the submarine cable but no subsidy was considered. The events that changed the mind of the British government were military, more precisely a military disaster known in Britain as the Indian Mutiny and in India as the First War of Independence. The mutiny/war of independence

[10] There were actually some hostilities between Britain and Vichy France during the Second World War but, I think, those can be disregarded.

EFFECT OF THE SUBMARINE TELEGRAPH; OR, PEACE AND GOOD-WILL
BETWEEN ENGLAND AND FRANCE.

Fig. 4.12

started in Lucknow on 10 May 1857. The report requesting help went by
telegraph from Lucknow to Calcutta, by traditional overland transport
to Bombay, by boat to Suez, overland to Alexandria, and again by boat
to Trieste, the nearest telegraph station. The message from Lucknow to
Trieste travelled for 40 days. A telegraphic connection to India had suddenly acquired priority.

The Red Sea and India Telegraph Company was founded in early 1858.
They asked the government for a dividend guarantee. Although the
Indian Mutiny was over by this time the government, in the new mood,
was willing to guarantee for 50 years a dividend of 4.5 per cent on capital of £800,000. The project was a disaster. Although each section of
the line worked at some time (as stipulated in the contract), they never
worked at the same time. Not a single message ever travelled along the
whole line between Bombay and Suez. It was an expensive failure for
the government. They paid the 4.5 per cent dividend to the company for
50 years.

Commercially, the most important enterprise was the Atlantic cable
but it was also of interest to the government. They offered a subsidy of
£14,000 per year during its working life, technical help by the Royal
Navy and the loan of warships.[11] The first attempt was made in August
1857. It had to be abandoned because the cable broke. The second
attempt started in the spring of 1858. There were a number of near-
disasters but the effort was finally crowned with success in August 1858.

[11] By a curious coincidence the name of
the battleship laying the first cable was
Agamemnon, the title of the first play in
which beacons are mentioned.

For the first time telegraphic communication was established between North America and Europe.

There was noisy rejoicing in the United States with salvoes of artillery and torchlight processions (one of which, unfortunately, set the Town Hall in New York on fire). The celebrations in England were more muted. There was a leader in *The Times* remarking that:

since the discovery of Columbus, nothing has been done in any degree comparable to the vast enlargement which has thus been given to the sphere of human activity.

The event was also commemorated in ponderous verse:

Two mighty lands have shaken hands
Across the deep wide sea;
The world looks forward with new hope
Of better times to be;
For, from our rocky headlands,
Unto the distant West,
Have sped the messages of love
From kind Old England's breast.

Also, as may be expected, there were cartoons in *Punch*. In the one shown (Fig. 4.13) the Anglo-Saxon twins (John Bull and Jonathan, certainly not Uncle Sam) jointly overturn the boats of despots.

Unfortunately, the celebrations did not last long. Queen Victoria and President Buchanan managed to exchange greetings, but the cable quickly deteriorated and stopped working altogether a couple of months later.[12] The sum total of messages passed was 732, some of them quite important. The British government is said to have saved £50,000 by countermanding an order to send two regiments from Canada to India.

The failure of the Atlantic Telegraph was no doubt a blow to the cable industry. The situation did not look rosy at all. By 1861 the total length of cables laid all over the world was 17,700 km of which only 4,800 km worked. The nineteenth century was, however, not the time for man to despair. A project was not abandoned just because it failed a few times. The third attempt to span the Atlantic was made in 1865. Unfortunately, the cable broke after 2,000 km was laid. The fourth attempt a year later succeeded however. In addition, the 1865 cable was hooked and repaired, so the end of 1866 saw both cables working. There have been uninterrupted communications between Britain and the US ever since.

The delay of seven years between the second and third attempts was caused partly by an investigation trying to find out what went wrong and why, but mainly because between 1861 and 1865 the Americans were otherwise engaged. They fought their Civil War.

[12] Queen Victoria wrote a bit dryly: 'The Queen desires to congratulate the President upon the successful completion of this great international work.' To which President Buchanan replied with more enthusiasm: 'May the Atlantic Telegraph under the blessing of Heaven prove to be a bond of perpetual peace and friendship between the kindred nations, and an instrument designed by divine Providence to diffuse religion, civilisation, liberty and law throughout the world.'

THE ATLANTIC TELEGRAPH—A BAD LOOK OUT FOR DESPOTISM.

John Bull. " HOLD FAST, JONATHAN."   Jonathan. " ALL RIGHT, JOHNNY."

Fig. 4.13

Submarine Telegraphy: The Beginning of the Second Industrial Revolution?

The two main functions of science are to explain and to predict. All sciences start with the explaining stage but some of them graduate to the predicting stage: these are the hard sciences. There aren't many of them. They are restricted to physics and chemistry. Biochemistry and biology might one day also become hard sciences but then they will probably be indistinguishable from physics or chemistry.

Explaining is in words. Predicting can also be in words but *real* prediction (and that is the only kind we shall talk about in this section) is mathematical. One sets up a model of the problem or system that is of interest, one puts this model into mathematical form, one solves the equations, and, lo and behold, predictions can be provided. An example will make it clearer how one can proceed from explanation to prediction.

It has been observed, ever since *Homo sapiens* have been able to make observations, that the Sun revolves around the Earth. An early explanation maintained that it was Apollo who dragged the Sun with his cart across the firmament. A more detailed and somewhat different explanation was provided by Ptolemy, who gave charts for the motion of the Sun and the planets. A hypothesis that seemed simpler and which fitted the astronomical measurements better was provided by Copernicus. He proposed a picture in which the Earth and all the other planets revolved around the Sun. This was a brilliant hypothesis but still soft science.

THE BEGINNING OF THE SECOND INDUSTRIAL REVOLUTION?

Kepler went further. He gave certain mathematical relationships which the orbiting planets had to obey. Newton went even further. From a postulate that bodies attract each other and from another postulate that the change of momentum is proportional to force he could derive the laws governing the motion of heavenly bodies. Was this an explanation or a prediction? It was both, but the predicting power was limited to what people knew anyway, the motion of the planets.

The first prediction that was really a prediction in astronomy was that of Halley, a friend of Newton. He made some measurements of the orbit of a comet and then using Newton's equations he calculated its future motion. He said the comet would return in 72 years. And it did, although Halley did not live to see it. One example is usually enough but it would be a pity not to give a particularly striking second example of the power of Newton's equations. The problem was with the planet Uranus. It was straying from its expected orbit. So either Newton's equations were wrong or there was some reason for the stray behaviour. It was calculated by Le Verrier and Adams that the orbit of Uranus measured by astronomers for the previous 60 years would be consistent with Newton's theory provided another planet existed. In fact, they were able to say where that unknown planet should be found. Le Verrier asked two Berlin astronomers with gallic-sounding names (Galle and d'Arrest) to look for the planet and five days later on 23 February 1846 the planet, soon to be named Neptune, was found. A science that can make such predictions is a hard science. It would, of course, be easy to give modern examples of the predicting power of physics. Robert Oppenheimer and his team were, for example, able to predict that, given the right circumstances, an assembly of uranium atoms will explode.

If sciences can be classified as hard or soft, what can be said about engineering? Does a similar classification make sense? Is all engineering that is done without the help of mathematics 'soft'? Well, that is a possible definition and it would place all the engineering efforts up to the middle of the nineteenth century into that category. There were, of course, many ingenious inventions before the nineteenth century; like the wheel, the water mill, the windmill, and the steam engine, to name a few. But they were not based on mathematical modelling. They were the products of long, painful development. The take-off into the modern world started with the interaction between physics and engineering using the language of mathematics. This interaction has formed the modern world.

How did it all start? It started with telegraphy, in particular with the laying of the Atlantic cable to which I shall now return in a little more detail. Everyone was aware of the seriousness of the problems involved but they were mostly traditional engineering problems to be solved by relying on traditional engineering skills, e.g. the paying-out apparatus that released the cable had to be flexible enough so that the extra stress resulting from the movement of the ship did not break the cable. And even problems like the safe storage of well over 1,000 miles of cable in a ship were far from trivial. However, there was considerable optimism that all these problems could be solved. Charles Bright, the engineer in charge of the whole enterprise, was enthusiastic and highly optimistic.

The mechanical problems were daunting but familiar. The electrical problems were entirely new and their seriousness was hardly appreciated in the beginning. It turned out that the current at the receiving end was not identical to the current at the transmitting end as illustrated in Fig. 4.14.

The first qualitative explanation (explanation in words rather than mathematics) of the way the current changed after it was switched on was given by Werner von Siemens, based on an analogy between the plates of a capacitor and the conductors in a cable. He claimed that the two conductors in a cable constitute a capacitor which needs to be charged up by the current flowing from the battery. Hence when the current starts to flow at the transmitting end some of the charge carriers are diverted to charge up this giant capacitor and that therefore they cannot reach the other end of the cable. Faraday, independently, came to the same conclusion.

So the cause of the change in current between the transmitting and receiving ends was identified but it was still not known how much the capacitance contributed to the deterioration of the signal and what the role of the resistance was. The economic importance of the problem

Fig. 4.14 The current as measured at opposite ends of a long telegraph wire.

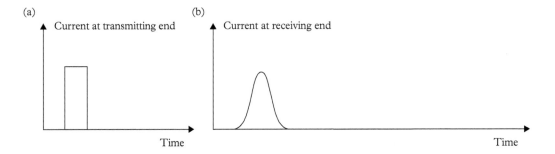

Fig. 4.15 Same as Fig. 4.14 but for a longer line.

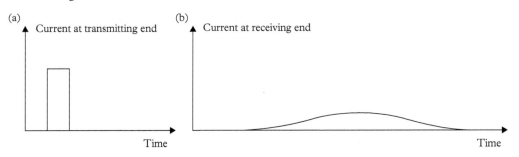

can be appreciated by looking at some further experimental results for even longer cables. The current at the receiving end then became even more spread out (Fig. 4.15b). Therefore one had to wait until this broad current form arrived before the next signal could be sent. Even worse, if a dot and dash were sent in succession (Fig. 4.16a) the received current form might have looked like the one shown in Fig. 4.16b. Now one can't even be sure whether one or two signals were sent, let alone whether they were dots or dashes. The inevitable conclusion was that the rate of signalling is bound to be considerably slower on long cables than on short ones. If this rate was below about six words per minute, then the Atlantic cable was not an economic proposition. The enterprise would not make a profit, no investor would be interested, and the project would never get off the ground.

This was the stage reached when William Thomson (later elevated to the peerage as Lord Kelvin) entered the fray. After some very sophisticated mathematical analysis he came to the conclusion that retardation (the time one had to wait before the next signal could be sent) was proportional to the square of the length of the cable. If a cable with a given length gave a retardation of one tenth of a second then a cable twice as long would give a retardation of four times as much, that is, four tenths of a second, and a cable 10 times as long would give a retardation of 10 seconds. This relationship became known as the 'law of squares'. It also followed from Thomson's calculations that the retardation for overhead lines was much smaller than for submarine or buried cables because the conductors in overhead lines were much farther from each other than those in the cables.

The kind of calculation William Thomson did was the first shot fired in the Second Industrial Revolution although no one was aware of it at the time. Calculations had, of course, been done before but they were relatively simple and the reasoning behind them could be understood by those who made the decisions. With Thomson's calculations this was no longer the case. None of those who were concerned with the technical aspects of the cable nor any of those who sat on the board of directors of the Atlantic Telegraph Company could understand any part of Thomson's reasoning. It was not their fault. It happened that for the

Fig. 4.16 The current measured at opposite ends of a long telegraph wire when a short and a long pulse are sent.

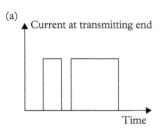

(a)

Current at transmitting end

Time

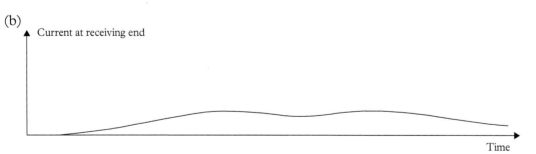

(b)

Current at receiving end

Time

first time in history pure science intervened in matters industrial and the directors could not possibly perceive that some incomprehensible scribbling on a few sheets of paper had the capability of predicting the future performance of the cable. After all, simple experiments followed by simple calculations always sufficed in the past. Why should one suddenly come to the conclusion that a professor of natural philosophy at the University of Glasgow knows better how to solve the problem than the electricians on the job?

Why make theoretical calculations at all in a field of practical significance? It might be necessary to do such things for the motion of heavenly bodies which do not easily lend themselves to experimentation. Theory can explain why Uranus strayed from its orbit but surely in matters of practical significance conclusions have to be drawn from carefully executed measurements. The sensible thing to do is to manufacture the cable, wind it on lots of big coils, put the transmitting apparatus at one end and the receiving apparatus at the other end, and send signals along the whole length of the cable. If the rate of transmission of signals is fast enough the cable is OK. If the rate of transmission is not fast enough, then the cable must be modified. Surely, by experimentation alone one would eventually arrive at the correct design. This argument is fine but there is a snag. Each experiment of that kind would have cost several hundred thousand pounds, clearly not in the realm of practical possibilities. The decision about the dimensions of the cable to be produced and about the manufacturing process to be followed had to be made in the absence of sufficient experimental evidence. But could one make some relevant experiments by joining together a few of the existing cables which joined Britain to the Continent? That's exactly what E. O. W. Whitehouse, the man who was later in charge of all the electrical engineering problems in the Atlantic cable project, did. He conducted such experiments in 1855, measured the broadening of the pulses, and concluded that Thomson's predictions were wrong. He gave a lecture at the British Association Meeting in Cheltenham in 1856 entitled 'The Law of the Squares—Is it Applicable or Not to the Transmission of Signals in Submarine Circuits'. He concluded:

I believe nature knows no such application of that law and I can only regard it as a fiction of the schools, a forced and violent adaptation of a principle in Physics, good and true under other circumstances, but misapplied here.

Whitehouse felt that he should not go too far in his criticism. The professor's calculations were no doubt correct—somewhere where the principles of physics played a role—but not in this particular practical problem. The expression, 'a fiction of the schools' made it clear what he thought of Thomson's theory.

What was the argument about? It was about design of the cable: what the diameter of the conductor should be and how thick the insulator between the conductors should be. The main disagreement was on the diameter of the conductor. According to Whitehouse the conductor could be thin, comparable with those used in previous cables. Thomson favoured thicker cables because that was suggested by his calculations. The board was unanimously in favour of a cable as thin as possible. The reasons were clear enough. The thinner the conductor, the cheaper it was. The final decision was in favour of a cable that Thomson could accept.

It may be worth mentioning at this stage that Thomson was one of the directors of the Atlantic Telegraph Company. One might be permitted to think that he was elected on the grounds of his superior knowledge of electricity and mathematics, but that assumption would not be borne out by the facts. He was elected a director by the Scottish shareholders as an eminent man who would be able to represent their interests on the board. Did the board of directors want to involve Thomson in the electrical problems of the cable? Sometimes they grudgingly agreed to consult him, but on the whole they did not think that Thomson's science was relevant to the working of the cable. As it happened, Thomson followed the progress of both attempts in 1857 and in 1858, sailing with the *Agamemnon*, but not in an official capacity. In the words of S. P. Thompson, an early biographer of William Thomson, 'The work which he undertook for it was enormous; the sacrifices he made for it were great. The pecuniary reward was ridiculously small.'

One could write several chapters about the contributions of William Thomson to the cause of the telegraph. I mention here two more of those contributions. He insisted, as a director of the company, that the manufacturers of the cable should work to stringent specifications, and he set up the first factory control laboratory that ever existed. Secondly, the device which made it possible to receive signals across the Atlantic (Whitehouse's own apparatus turned out to not be sufficiently sensitive to receive the weak signals) was Thomson's own invention, a mirror galvanometer (similar in principle to that of Gauss and Weber) adapted specially for working on a rolling ship. The last two contributions were quite crucial in being able to use the 1858 cable for the few months it worked, but they were not fundamental. They could have been done by lesser men. Perhaps not in 1858 but certainly within a decade or so.

I contended earlier that the Second Industrial Revolution started with the work of William Thomson. For that, we need a definition of the First and Second Industrial Revolutions, terms which have been used loosely in the past. My favoured definition is that the First Industrial Revolution replaced muscle power with machinery, and the Second Industrial Revolution, which is still running in our time, is in the process of eliminating the need for brain power.

There is no doubt that scientists did make significant contributions to the development of the machines fuelling the First Industrial

Revolution (e.g. Professor Black, also from Glasgow) but the technologists would have managed on their own. The Second Industrial Revolution is in a quite different category. It came about as a result of a unique blend of interaction between science and technology with occasional government encouragement (the most expensive example being the Manhattan Project).

William Thomson was the pioneer in producing the theoretical work crucial for the progress of communications but very soon Maxwell and his disciples took over. I shall discuss their contributions later when coming to wireless telegraphy. It might, however, be instructive to ask at this stage the question: how important was the work of the leading scientists in the hundred years from 1815 to 1914? What would have happened if three or four dozen of those scientists had died in their infancy? There is no doubt that our lives would be entirely different. We would have no telephone, no radio, no television, no computers, no X-ray diagnosis, and no magnetic resonance imaging.[13] One may very well contrast this with the influence of great politicians. If in that period all the rulers of Europe and all their prime ministers had perished before they assumed office, that would have made only very minor differences in the world. Perhaps the Franco-Prussian war would never have taken place and the First World War might have been fought between different alliances but that is probably all. Those eminent scientists would have made the same discoveries and inventions whoever had been the rulers.

Britannia Rules the Cables

As we have seen, Britain was first in the world to establish a commercial telegraph and the first in the world to establish a submarine cable industry. Did other countries catch up? In setting up telegraph lines, yes, very soon. In establishing a submarine cable industry, no. It was an expensive business and the expected returns were small. When any of the European nations wanted to set up lines under the seas, more often than not they entrusted the project to a British company. It was a British company, for example, which established the first reliable link (a number of previous attempts had failed) between France and Algiers in 1870. The British dominance was noted by William Siemens in his 1873 presidential address to the Society of Telegraph Engineers:[14]

London...is the principal centre of the Telegraphic enterprise in the world, and musters consequently the greatest number of Telegraph Engineers. It is a remarkable fact that the manufacture of insulated wire, and of submarine cables, is almost entirely confined to the banks of the Thames.

The wiring up of the seas and oceans followed soon, covering practically the whole world by the end of the century.[15] This was clearly what Puck of *A Midsummer Night's Dream* fame had in mind when he

[13] No hydrogen bomb either, but that's a separate question.

[14] Born as Karl Wilhelm, he was one of the younger brothers of Werner von Siemens. He emigrated to England in the 1840s, became naturalized in 1859, was elected a Fellow of the Royal Society in 1862, and was knighted shortly before his death in 1883.

[15] It may be worth mentioning here a late arrival (1902) to the network which connected New Zealand, Australia, and Canada. Its fame is partly due to the fact that it broke the monopoly of Eastern Telegraph but it also induced Rudyard Kipling to celebrate the event in verse far superior to that written by other telegraph enthusiasts:

Here in the womb of the world, here on the ribs of earth,
  Words and the words of men, flicker and flutter and beat—
Warning, sorrow and pain, salutation and mirth—
  For a Power troubles the Still that has neither voice nor feet.

promised 'to put a girdle round about the earth'. Although his original estimate of forty minutes for doing the job was far too optimistic, the actual speed of laying the cables may still be regarded as impressive. The growth in the length of submarine cables in the period 1865 to 1908 is shown in Fig. 4.17. As may be seen, most of it was owned by private industry. British dominance was overwhelming. In 1896 as much as 87 per cent of submarine cable was manufactured in Britain. Out of a total of 30 cable-laying ships 24 were British. For the ownership of cables in the years 1892 and 1908 see the pie charts of Figs. 4.18a and 4.18b. Although Britain's share kept on declining it was still 56 per cent of the world total in 1908.

Was the British ownership of the cables an advantage to Britain? Yes, it was a tremendous advantage whenever conflicts arose. The possible theatres of conflict were the colonies, and the power with which Britain came most often into conflict was France. The British Foreign Office regularly intercepted telegraphs concerned with French colonial matters and then used the information in Britain's best interests.

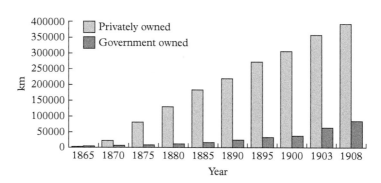

Fig. 4.17 The total length of cables in service in the world in the period 1865 to 1908.

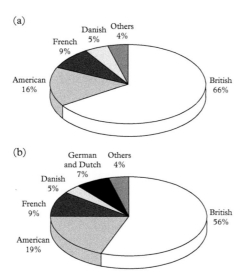

Fig. 4.18 Distribution of cables by ownership in (a) 1892 and (b) 1908.

In 1885 the news of the defeat of a French fleet at Tonkin was relayed to the British ambassador in Paris before the French government heard of it. In 1893 the French government's authorization to Admiral Humann to issue an ultimatum to Siam was read by the British Cabinet well before the admiral received it. Perhaps the events in Morocco in 1894 were even more characteristic of the weight of British power. When the sultan died, the country's only telegraph line from Tangier was requisitioned by the British consul in order to communicate with the Foreign Office. Havas, the French news agency, reported afterwards from Madrid:

Newspapers complain that the English cable, the only one working between Tangier and Europe after the break in the Spanish cable, was seized during the whole of last night by the English minister to communicate with the Foreign Office.

The newspapers wonder what can be, under these conditions, the security of the interests of other nations if England, holding all sources of information, can thus suspend at will communications which are not its own.

In 1895 the news that French forces occupied Tananarive, the capital of Madagascar, was held up in London for three days. The French were not pleased. In fact, by this time, the Germans were not pleased either. In January 1896, when the Boers successfully repulsed a raid carried out by Jameson, an agent of Cecil Rhodes, with the tacit approval of the British government, the German emperor sent a congratulatory telegram to President Kruger of the Transvaal. The telegram was leaked to the English newspapers before it reached the president. It rose to fame as the Kruger telegram, ending the harmonious relationship between Germany and Britain.

The Boer War broke out in 1899. Communications in cipher between South Africa and Europe were subsequently banned by the British government and there was serious interference, even with diplomatic traffic. In fact, in January 1900, the Home Office issued a warrant to the Post Office

to produce, for the Information of the Intelligence Department of the War Office, until further notice any telegrams passing through the Central Telegraph Office (in London), which there is reason to believe are sent with the object of aiding, abetting, or assisting the South African Republic and the Orange Free State.

Did France and Germany do anything to undermine British dominance? In July 1886 the French government envisaged a project for a French cable along several routes, which would have included the setting up of cable factories and would have needed a subsidy of a million francs (£40,000). The Chamber of Deputies, after lengthy considerations, rejected the bill a year later. They thought it was too expensive. Not until the Fashoda incident in 1898 (to be discussed later) did the

French take a serious interest in cables. Their expenditure on cables between 1900 and 1914 was close to £300,000 per year. They served a strategic purpose but they never paid for themselves. The Germans made similar efforts. They fared somewhat better. Their transatlantic line via the Azores was an immediate commercial success but, on the whole, their reasons for building cables were also strategic.

Did France and Germany combine forces to combat British hegemony? Oddly enough, they did to quite a large extent. It was odd because the period of cooperation between the telegraph authorities of the two countries coincided with frosty relations between their foreign ministries, domiciled at Quai d'Orsay and at Wilhelmstrasse. Both the French telegraph authorities and the French public disliked the necessity of relying on British cables, in spite of the *entente cordiale* signed between France and Britain in 1904. A cartoon (Fig. 4.19) depicts John Bull as a malevolent coachman riding all the cables leading to places within the French sphere of interest: Martinique, Madagascar, Algiers, Tunis, Morocco, and Tonkin.

Was there any conflict between the United States and Britain? Not initially. The Americans were busy establishing communications across their continent. They had no colonies. They needed no cables. But all that was changed by the Spanish–American War of 1898. When the Americans won the war (not that the outcome was in doubt for any moment) they suddenly found themselves in charge of the Philippines and of a number of assorted islands thousands of miles away from the mainland. In their new imperial mood the Americans also persuaded themselves to annex Hawaii in 1899, which by that time was not particularly independent anyway. The problems of communication with

Fig. 4.19 The English coachman holds all the reins to places in the French sphere of influence.

overseas properties suddenly presented themselves. In 1899 President McKinley said:

The necessity for speedy communication between the United States and all these Pacific islands has become imperative. Such communications should be established in such a way as to be wholly under the control of the United States, whether in time of peace or of war.

Quickly, a large number of bills were introduced in Congress with the aim of setting up cable enterprises with various amounts of government subsidy. Congress refused to choose between the various schemes. They rejected them all. They had not yet realized that the three things needed to run an empire are money, money, and money.

In 1901, however, America entered the fray. The all-American Pacific Cable Company was founded, whose all-American cable reached Manila in July 1903. Later it turned out that they were not quite as all-American as was first assumed. Secret deals ensured that one half of the American Pacific Cable was owned by the Scotsman John Pender's Eastern Telegraph and a further quarter by the Great Northern Company, a Danish firm.

So why was it that Britain could keep her hegemony so successfully? The first reason was because the British were there first. They had the machines, the men, the materials, the ships, and the organization. The second reason was the existence of the British Empire. Only the British needed a truly worldwide network, and only the British could generate the amount of commercial activity which made the cables pay. There was very little subsidy, nearly everything was done by private capital. It would be, however, naive to assume that there were no contacts between Eastern Telegraph, the dominant cable company, and the British government. Many of the company's directors got their jobs more for their abilities to curry government favour than for their commercial acumen. The dependence was mutual. In her hour of need the British Empire could count on Eastern Telegraph. Conversely, had the existence, or just the worldwide dominance, of Eastern Telegraph been threatened, Britain would have surely stepped in. This last argument is probably the most potent one. The telegraph industries of other countries in receipt of vast government subsidies would have surely been able significantly to reduce the influence of Eastern Telegraph. But had they made a serious attempt—ultimately—they would have had to face the might of Britain. The situation may be characterized by a maxim of possibly more general validity: 'it is not relevant how ruthless a power actually is; what matters is how ruthless that power might eventually turn out to be'.

Ruthlessness and speed were the keywords when the First World War started. Within four hours after the British ultimatum to Germany expired, a British cable ship managed to cut all five German cables: two

to the Azores and North America, one to Vigo, one to Tenerife, and one
to Brest.

The end of the First World War found Europe poorer and the US
richer. Britain still owned as much as 50.5 per cent of the world's cables
in 1923 but the writing was already on the wall. British influence kept on
declining for two reasons: partly because technological advances
brought about a new family of cables and partly because of the appearance of new media for communicating, the telephone and the radio.
I shall come back to these problems in the following chapters.

Morse Code versus Binary Coding

I have already mentioned the binary code, the codebooks introduced by
Claude Chappe, Morse code, and that of Gauss and Weber. The choice of
dots and dashes by Morse turned out to be particularly advantageous
when the human ear was used for detection. The best operators achieved
transmission rates of up to 40 words per minute.

One might be allowed to think that Morse code is a binary code
because only two symbols are used, dots and dashes, but this is not
counting the spaces which are needed between symbols, between letters, and between words. The two codes are indeed quite different as an
example will show. How long would it take to transmit the word 'eat' by
one or the other code? The Morse codes for letters e, a, and t are $\cdot$, $\cdot$ $-$,
and $-$, respectively. Hence one needs to transmit

e		a		t
dot	longer space	dot-space-dash	longer space	dash

Assume now that a dot is one unit of time, a dash is two units, a space
between symbols is one unit, and a space between letters is 3 units. Then
the total length of transmission is 13 units:

E		A		t
1	3	1 1 2	3	2

As a digression it may be worth mentioning here that Morse code
could be and has been used with combinations besides dots and dashes,
for example, by positive and negative currents. This can be put into
practice with relative ease by switching the battery.

Let us now look at the length of the message using a binary code in
which the 26 letters of the alphabet are coded by five digits (remember
that was the code invented by Cardano in the sixteenth century). How
long would it then take to transmit the word 'eat' in a binary code? Each
letter would need five symbols (0 or 1) so the total time to transmit the
three-letter word is 15 units. Would spaces be required between the letters? No, there is no need for a space between the letters because each
letter is of the same duration so there is no need to give a warning when

a letter ends and a new letter starts. There is *always* a new letter after every five units.

In this example Morse code happens to give a shorter transmission time than the binary code but the reason is simply that a word very favourable for Morse code has been chosen. Had the word 'joy' been taken the result would have been quite different. The Morse codes for the letters are then · − − −, − − −, and − · − −. By similar calculation, the total length would now come to 33 units of time. With the binary code it would still be 15 units. On average, the binary code is faster than Morse code by about 50 per cent.

If it's Tuesday, it Must be Belgium

There used to be some Americans (and they might not be entirely extinct yet) who had difficulties in distinguishing European countries from each other. So when they did their European tour how could they find out which country they were in? Very simply, by reference to a fixed point in time. For example's sake let us assume that they arrived in Stockholm on Friday morning and according to their itinerary they were to stay in Copenhagen on Saturday, in Hamburg on Sunday, in Amsterdam on Monday, in Brussels on Tuesday, in Paris on Wednesday, and in Madrid on Thursday. The crucial information was the relationship between the days and the countries. If the time was Tuesday lunchtime, then the country had to Belgium.

One could imagine a similar correlation between letters of the alphabet and time slots, which for simplicity will be taken one second apart. If an electric pulse arrives five seconds after the agreed starting time, then it must be E; if it comes eleven seconds after the starting time then it must be K. Clearly, there is no need to code the letters. All that is needed is a reference time and a rotating wheel upon which the letters of the alphabet are embossed. When the electric pulse arrives, the wheel is pressed against a strip of paper and the relevant letter is printed. Such a machine was invented by David Hughes in 1854.

Time Division Multiplex

The title sounds a little intimidating, particularly the word multiplex. It must be accepted as part of the current jargon used by communications engineers. The idea is really simple and strongly related to the time slots mentioned in the previous section. The aim is for several (a number N) operators to send messages simultaneously on a single line. For example, if there are only 3 operators ($N = 3$) and they wish to transmit the words 'eat', 'joy', and 'dog', respectively, then they use the following method. The first operator puts his first letter (or rather its five digit code) e on the line. This is followed by the second operator putting his first letter *j* on the line, and then the third operator putting *d* on the line. After that comes the second letter of the first operator, etc. Thus the three words are carved up and transmitted as follows:

e   j   d   a   o   o   t   y   g

Every third letter belongs to the same operator. The hardware that can do this is less simple. There must be a way of storing letters until they come on the line (the operator cannot be expected to be so precise in his timing as to put in his signal at the exact time when the previous operator finishes his) and there must be synchronization between the transmitting and the receiving end so that every Nth letter goes to the same operator. The advantage of the system is that each operator can work with his normal speed but the overall transmission is N times faster.

Such a system was invented by Emile Baudot, a French engineer.[16] It went into service in France in 1877. The telegraphic services of most countries adopted similar devices afterwards. The significance of this invention was not only to speed up transmission but to establish a new principle which has come into its own with modern digital techniques (see Chapter 11).

Diplomacy

Diplomacy is the art of the possible. In order to know what is possible and what is not, one has to give due consideration to the matter. Giving due consideration takes time. If a message is in transit for 30 days and the reply will also travel for 30 days, then the recipient may safely assume that he has a good week to chew matters over. On the other hand if messages take no more than a few minutes to arrive, then decisions might be required in comparable time. It seems therefore unlikely that many of the nineteenth-century diplomats did actually welcome the advent of fast communications. We have on record the words of Edmond Hammond, permanent under-secretary to the Foreign Office from 1854 to 1873. He complained that the telegraph tended 'to make every person in a hurry, and I do not know that with our business it is very desirable that it should be so', for it tempted officials to 'answer offhand points which had much better be considered'. Their comfortable work must have been rudely interfered with.

For some other diplomats the telegraph might have saved the trouble of thinking. Sir Horace Rumbold, British ambassador to Vienna, deplored 'the telegraphic demoralization of those who formerly had to act for themselves and are now content to be at the end of the wire'.

At the beginning the telegraph was used in the diplomatic service for obtaining and sending on routine pieces of information. Until 1859 diplomatic traffic in Britain had the same treatment as the rest. A telegram

[16] The unit of the signalling rate used by communications engineers is called a baud in his honour.

addressed to the Foreign Office arrived at the nearest telegraph office from where a messenger walked over to Whitehall and handed over the telegram to some lowly official. Similarly, when the Foreign Office wanted to send a telegram the lowly official walked over to the nearest telegraph office where, presumably, he jumped the queue if there was one, and presented the piece of paper to the operator. After 1859 the Foreign Office went as far as to hire two clerks to accept telegrams sent after normal working hours. In a few cases, however, the telegraph did already play a significant role as, for example, in 1856 at the Congress of Paris where the negotiations ending the Crimean War were conducted. The British representative was Lord Clarendon, Foreign Secretary in the first Palmerston Government. The prime minister sent his instructions by coded telegrams to the Foreign Secretary.

By the 1870s the telegraph became all-pervasive. The more important government offices had direct telegraph lines and soon the Foreign and Colonial Secretaries could not even evade the telegraph by leaving the office. Direct lines were set up both to their London homes and to their country estates.

Even with hindsight it is difficult to decide whether fast communications have made wars more likely or less likely. In principle, the possibility of quickly ironing out misunderstandings should have made the telegraph an agent of peace, but no near-wars are on record which were avoided thanks to the telegraph, unless one includes an Anglo-French confrontation in 1898 at Fashoda in the Sudan. In that particular case war was indeed averted, but the best conclusion one could draw from the affair is that 'the power that has a telegraph at its disposal is better off than the power that hasn't.' The events are known as the 'Fashoda crisis'. The role the telegraph played will be discussed later.

Which other telegrams were consequential? Apart from the Kruger telegram that has already been mentioned, there was the 'Ems telegram', which preceded the French declaration of war upon Germany in 1870, and the 'Zimmermann telegram', which precipitated the entry of the United States into the First World War, and it is worth adding a further telegraphic exchange between two powerful men in Russia in 1917 which might have, possibly, prevented the Bolsheviks taking power.

Let me start with the events which led to the Franco-German war in 1870. They started in Spain two years earlier when a *coup d'état* by a radical general deposed the ruling monarch, Isabel II. Her long reign took place in a period which may be said to have been turbulent owing to a *coup d'état* every couple of years or so. Since elections were always rigged in favour of the ruling party the popular method for change of government was a *pronunciamiento* (the Spanish name for a type of *coup d'état*). The one that occurred in 1868 was somewhat different because it ended Isabel's rule. The reason was not her policies—she never pretended to have any—it was her choice of lovers which led eventually to her demise.

Having got rid of Isabel II the Spaniards decided that after all they did want a monarch. Unfortunately, in the circumstances it was difficult to find anyone even mildly interested in the job. Those to whom the crown was offered politely but firmly declined. In desperation, the Spaniards approached Leopold, who belonged to the Catholic branch of the Hohenzollern family. He thought he might have a go, and said so to the Spanish delegate on 19 June 1870. And that is where the telegraph comes into the story for the first time. The delegate sent a coded telegram to Madrid saying that he would return by 26 June and present the results of his negotiations to the Cortes, which was to remain in session until 1 July. Unfortunately, owing to an error by a cipher-clerk the date received was 9 July. Clearly, it would have been too much to ask members of the Cortes to brave the oppressive heat of Madrid until 9 July so the session was prorogued until November. When the delegate arrived on 26 June and found the Parliament building deserted he had to give a reason for the recall of the Cortes. The whole thing came out into the open. There was then no chance of presenting the French with a fait accompli.

The news reached Paris on 3 July. The French military in the court of Napoleon III had a look at the map and discovered that France was threatened by Hohenzollerns on all sides. The emperor of the French demanded that the king of Prussia, Wilhelm I, should persuade his distant relative to renounce his candidature. Wilhelm had for long been of the opinion that the Spanish throne was more bother than it was worth so he was quite willing to use his influence. Leopold, never too keen to take the job, was ready to call the whole thing off. On 12 July, Leopold's father renounced the throne on behalf of his son. That would have been the end of the story but for the ambitions of Napoleon III and of the Bonapartists in the court. They wanted to humiliate Prussia. The French ambassador was instructed to demand that Wilhelm I should give an assurance that Leopold's candidature would never be renewed. That was a bit too much for the king to swallow. He said, no.

With a bit of luck the crisis could have still been resolved had the normal procedures of diplomacy, valid in pre-telegraphy times, been followed. The king, taking the waters at the summer resort of Ems, would have then sent a letter to Bismarck in Berlin who presumably would have travelled to Ems and after an audience with the king would have drafted a reply to the French. Since the King was 73 years old and did not hanker after military glory, it is quite likely that the message reaching the French would have been worded with care to avoid an immediate confrontation. As it happened, the king sent a long, rambling telegram from Ems to Berlin describing his meetings with the French ambassador. At the end of the telegram he gave free hand to Bismarck to communicate what had been said to the ambassadors and to the press.

Bismarck first consulted the military. When he was assured that they were ready for war without any delay he had a go at the telegram. He did not change a single sentence. He just made it a lot shorter. The text that the world got to know was rather terse:

After the news of the renunciation of the Prince of Hohenzollern had been officially communicated to the imperial government of France by the royal government of Spain, the French ambassador at Ems further demanded of his Majesty the King that he would authorize him to telegraph to Paris that his Majesty the King bound himself for all future time never again to give his consent if the Hohenzollerns should renew their candidature. His Majesty the King thereupon decided not to receive the French ambassador again, and sent to tell him through the aide-de-camp on duty that his Majesty had nothing further to communicate to the ambassador.

The French recognized a snub when they saw one. They did not even wait for their ambassador to return from Ems. They declared war on 15 July. Seven weeks later the war was over. The principal French Army and Napoleon III himself surrendered at Sedan.

In spite of her victory Germany remained, in the words of Bismarck, a 'saturated' power with no ambition of her own that could lead to war. The situation changed, however, in 1890 when Bismarck was dismissed by the young Kaiser, Wilhelm II. Germany started to flex her muscles. Britain's domination of the world suddenly became threatened by a new power. The first indication of such a threat came with the so-called Kruger telegram sent by the Kaiser to the president of the Transvaal. The political background has already been mentioned. The Jameson raid was repulsed by the Boers in the first few days of the year 1896. The British aim of acquiring the rich mines of Johannesburg suffered a set-back. The Kaiser made no secret of his sympathies as the text of the telegram shows:

I sincerely congratulate you that, without appealing for the help of friendly Powers, you with your people, by your own energy against the armed hordes which as disturbers of the peace broke into your country, have succeeded in re-establishing peace and maintaining the independence of your country against attacks from without.

So the Germans had started to be troublesome—a bit unexpected from an old ally and from one that was ruled by a grandson of Queen Victoria. At the same time conflict was also looming with France, Britain's traditional enemy.

Having lost the war of 1870–1 France looked for consolation in colonial expansion. The Fashoda crisis came about because the ambitions of France and Britain in Africa could not be simultaneously satisfied. In the last quarter of the nineteenth century both powers pushed forward the borders of their colonies with unseemly haste. The great British dream was to have a broad red stripe on the map stretching from Cairo to Cape Town along the eastern coast of Africa. The French counter-dream was a continuous blue swath from Dakar on the Atlantic to Djibouti on the Red Sea. According to the rules of geometry at least one of the two powers was to fail in her ambition.

The French champion was Major Marchand, whereas the colours of Britain were hoisted by General Kitchener. Marchand started his journey from the French Congo in West Africa in January 1897. The journey took quite some time. He arrived at Fashoda on the Nile on 6 July 1898 where he speedily unfurled the French flag. For the British, the journey would have taken much less time since the Nile was navigable to Fashoda. Their difficulty was not the terrain and the elements but rather the Mahdi, and his followers the dervishes, who some 17 years earlier had overrun the Sudan, taken Khartoum, and killed General Gordon. Kitchener came with an army of some 26,000 men. He defeated the dervishes at Omdurman on 24 August. He did not stop for long to celebrate his victory. By 10 September he was ready to sail to Fashoda, 500 miles up the Nile. He arrived with his five gunboats nine days later, only to find Marchand in possession of the old fort.

The meeting between Marchand and Kitchener was not entirely cordial. Neither of them was willing to yield but at least they observed the formalities of civilized behaviour. Some lukewarm whisky was offered to Marchand who accepted it but as he remarked afterwards, drinking '*cet affreux alcool enfumé*' was one of the greatest sacrifices he had ever undertaken in the service of his country. At the end they agreed to disagree and let their respective governments sort things out. Kitchener planted the Egyptian flag, left a battalion of Sudanese infantry behind, and sailed back to Omdurman from where he had a telegraphic line to London. He described Marchand's predicament in vivid terms:

He is short of ammunition and supplies... he is cut off from the interior and his water transport is quite inadequate. He has no following in the country and had we been a fortnight later in crushing the Khalifa nothing could have saved him and his expedition from being annihilated by the dervishes. Marchand quite realises the futility of all their efforts and he seems quite as anxious to return as we are to facilitate his departure.

Soldiers are usually pictured as blunt fellows who are in the habit of blurting out the truth. This particular telegram, and those following it up, show Kitchener as a master of deception of the same rank as the finest diplomats. Not a single statement concerning Marchand was true. He had plenty of ammunition, an excess of tinned food, and even his own vegetable garden. He had better transport facilities than the garrison left behind by Kitchener. He had already signed treaties with the natives. He had already had one encounter with the dervishes before Kitchener's arrival and had forced them to withdraw. He was steadfast in his mission and had no intention whatsoever of giving up his post at Fashoda.

But Marchand had no communications with France. The French government knew only Kitchener's version, but even that was enough to inflame passions. The popular press, and in fact not only the popular press, reacted angrily on both sides of the Channel. In *Le Petit Journal* France was depicted as Little Red Riding Hood threatened by the Big Bad Wolf of heavily armed Albion. The cartoonist of *Punch* drew Marchand as a quarrelsome Gallic cock being reminded of the spirit of Wellington by the British prime minister.

The stand of the British government was unequivocal. Withdraw from Fashoda, or war, was the message to the French government although not in exactly those words. In any case, just to put a little more pressure on the French, the Mediterranean fleet was alerted. Had Marchand had access to the French press and had he been able to send telegrams every day depicting in dramatic terms his stand against British aggression it seems very unlikely that public opinion would have allowed any French government to climb down. As it happened, they had some space to manoeuvre. After lengthy deliberations they climbed down. Marchand, in spite of his protests, was given instructions to withdraw. The crisis was over.

The telegraph did not seem to play any significant role in the outbreak of the First World War. Telegraph lines were, of course, buzzing all over Europe but there is no evidence in either direction as to whether it made the war more likely or less likely. It was probably neutral. One would have needed more than a telegraph, probably a miracle, to prevent the war breaking out. On the other hand the entry of the United States into the war was far from certain, even as late as 1916. The German high command certainly made a mistake when it got into the habit of sinking American ships. No neutral power likes to have its ships sunk. Thus, the likelihood of American intervention considerably increased after January 1917 when the Germans decided to step up their submarine campaign. There was, however, still vacillation. The isolationist lobby was still strong in the Senate. What precipitated the entry of the US into the war was a telegram sent in January 1917 by Zimmermann, the German Foreign Minister, to his ambassador in Mexico.

How could Germany communicate with Mexico during the First World War? Not through their own cables to America since, as mentioned previously, those were cut at the very beginning of the war. They could, though, communicate by courtesy of the Swedes,[17] directly by radio (as will be briefly discussed in Chapter 5), and, oddly enough, by using the American diplomatic cable from Berlin to Washington.[18] Everything would have been OK for the Germans but for the inquisitiveness of British intelligence who managed to break the German cipher. The text of the telegram was amazing. It started by telling the Mexican government of the German intention to start unrestricted submarine warfare, and then went on:

[17] The German diplomatic telegrams were first sent to Stockholm where they were re-enciphered by the Swedish Foreign Office and sent by British cable to Buenos Aires, whence the German ambassador retransmitted them to Washington, DC, through the American telegraph lines. The route was known as the Swedish roundabout.

[18] The permission was given by Woodrow Wilson, the president himself. After his re-election in November 1916 he made great efforts to end the war by some kind of compromise between the belligerents. By permitting the Germans to use the American diplomatic channel in cipher, he hoped to remove obstacles from the peace negotiations.

We make Mexico a proposal of alliance on the following basis: make war together, make peace together, generous financial support, and an understanding on our part that Mexico is to reconquer the lost territory in Texas, New Mexico, and Arizona

The ambassador was further instructed to contact the Mexican president if war broke out and to ask for his services to recruit Japan into the alliance. Japan's reward, although this was not in the actual telegram, would have been the state of California.

The telegram was released in America to a public partly horrified, partly incredulous. The pro-Entente lobby saw proof of German expansionism; the isolationists maintained that the telegram was a forgery. Senator O'Gorman, a man of pure Irish descent, recognized the machinations of perfidious Albion. Fortunately for the Entente, all speculations about authenticity ended with Zimmermann's admission at a press conference: 'I cannot deny it, it is true.' What Zimmermann's motives were in sending the telegram, and particularly in admitting authorship, has never been satisfactorily answered by historians. The likelihood is that German internal politics played a significant role. Whatever the motive was, the effect was a disaster for German foreign policy. American public opinion, mostly neutral up to that time, was incensed. The legislators could not ignore the public mood. On 2 April, less than five weeks after the publication of the telegram, the United States declared war on Germany. In the announcement the telegram was actually named.

The next jump in time is a modest one from April to August of the same year, 1917. The venue is Petrograd, the capital city of Russia.[19] The political situation, to say the least, was precarious. The events which were unfolding were of crucial importance for the history of the world for the rest of the century.

The changes started with the revolution in February. The tsar was forced to abdicate and a provisional government took power. The government took a number of liberal measures, but as usual in the wake of revolutions, the measures were too liberal for the right and not revolutionary enough for the left. A major bone of contention was the decision to continue the war against the Central Powers. It greatly displeased those who had been providing the cannon fodder for the previous three years. Military discipline declined, industrial anarchy increased, local incidents ending in bloodshed multiplied. By August, the two most important men in the country were Kerensky, the head of the provisional government, and Kornilov, the commander-in-chief of the armed forces.

Kornilov wanted the prestige of the officers raised, a logical demand at a time when some units got rid of their officers by the simple expedient of killing them. He would have been willing to put up with committees and commissars but wanted their responsibilities clearly

[19] It used to be St Petersburg. The name was changed to Petrograd in 1914, to Leningrad in 1924, and back to St Petersburg in 1991. The wheel came full circle.

defined. Kerensky, a socialist by party affiliation, recognized that the only way to stem the tide of anarchy was to implement the measures proposed by Kornilov even if it meant a break with the Soviets who represented the interests of the workers and peasants. He was ready to sign a string of measures aimed at improving discipline. That's when V. N. L'vov, procurator of the Holy Synod in a previous cabinet, appeared on the scene. He offered himself as an intermediary. He brought a message to Kerensky from Kornilov. Kerensky understood the message as Kornilov wanting to become the supreme dictator. Kornilov was under the impression that he had offered his full cooperation to Kerensky in case of a backlash from extremists, and had indicated his willingness to accept a post in Kerensky's cabinet. There was a misunderstanding.

Kerensky's next step was to seat himself at the telegraphic apparatus and contact Kornilov in Mogilev.[20] Had Russia been less backward they could have discussed the matter on the telephone. Had Russia been even more backward so as to have no telegraphic connections at all, they would have met in person and, quite likely, they would have hammered out their differences. Unfortunately, they had to rely on the telegraph. The main characteristic of messages sent by telegraph is their conciseness. So instead of asking the obvious: 'Now what is it exactly that you want?' and wait patiently until the lengthy explanation takes shape on the apparatus, Kerensky asked for a simple confirmation. His question to Kornilov was essentially: 'Would you confirm the message that you sent with L'vov?' Kornilov, trying to be brief, said yes, without enquiring what the message was that he was supposed to have sent.

The misunderstanding was perfect. Kerensky dismissed Kornilov who, believing that Kerensky was a hostage of the extremists, moved against the capital. In order to fend off the deposed commander-in-chief Kerensky permitted the arming of the Red Guards (the military wing of the Bolshevik Party) and agitation among the troops, which did indeed succeed in causing anarchy in Kornilov's camp. Kerensky might have been right in thinking that he had managed to avoid the frying pan. Unfortunately, he fell straight into the fire. Within two months the Bolsheviks, who gained most from the suppression of Kornilov, took power in a carefully organized *coup d'état*, known widely as the October Revolution.

Had Kerensky kept himself far from the telegraph and had he met Kornilov in person, Lenin might have become known to history only as a minor German agent, and Stalin's predilection for murdering people might have been confined to those few whom he could kill single-handedly.

War

[20] It was the Hughes apparatus mentioned earlier which transmitted the actual letters, not codes, so it did not need a skilled operator.

War is supposed to be the continuation of diplomacy by other means, but as far as the telegraph is concerned diplomacy and war are on different footings. For diplomacy, the telegraph had made quick negotiations possible between the powers in conflict. For war, it facilitated communications between various military units and between the field of operation and headquarters for each of the belligerents separately.

Since time immemorial the main reason for setting up communications systems has been to satisfy military requirements. The mechanical telegraph, the first communications system capable of transmitting any information, would never have come into existence but for the French revolutionary wars. The electrical telegraph was born in peace but that did not, of course, preclude its use by the military. The sole reason why military applications came a little late was the serious shortage of wars in that period.

The first major military exercise after the advent of telegraphy was the Crimean War in which an alliance of Turkey, France, and Britain faced the might of Russia. As soon as the war broke out in March 1854 the allied governments hastily laid (there was not time for private enterprise to act) an overland line from Bucharest, where the Austrian line ended, to the Black Sea port of Varna where the troops embarked for the Crimea.

The British commander-in-chief was Lord Raglan, who did have plenty of combat experience in the Peninsular War but the intervening 40 years of peace had adversely affected his military prowess. He also had the disconcerting habit of referring to the enemy as the French which in the circumstances was not the right thing to do. The war is mainly remembered in Britain for the hardships the ordinary soldiers had to endure during the winter, for Florence Nightingale, and for the Charge of the Light Brigade. The French tend to remember their victories commemorated in Paris by names like Place de l'Alma, Avenue de Malakoff, and Boulevard Sebastopol.

The British side conducted the war with moderate enthusiasm. The terrible casualties caused by neglect were reported in the newspapers and brought about the fall of the coalition government (of Peelites and Liberals, led by Lord Aberdeen). The next prime minister was Palmerston, who had a great reputation as a champion of British interests. The French leadership was less worried about casualties and more about *gloire*. For Louis Napoleon, who had only recently acquired the emperor's garb, the stakes were high. In case of failure, he had a dynasty to lose. The different approaches to the conduct of the war became manifest when in the following spring (seeing that the war would drag on because the Russians showed no inclination to surrender), a 340-mile submarine cable was laid by Newall and Company from Varna to the battle zone.

This was the first time in history that a ruler inclined to take direct military command could actually do so from the comfort of his palace.[21] Napoleon III did make good use of the opportunity. Believing strongly in the inheritance of military characteristics he showered the unfortunate

[21] Napoleon I also had a chance to direct military operations with the aid of the mechanical telegraph system which was good for keeping in touch but too slow for giving detailed instructions. In any case, Napoleon I had a marked preference for being on the battlefield himself.

French commander with advice, instructions, and suggestions. The British War Office was much more modest: it restricted itself to enquiries as to the health of Captain Jarvis, who was believed to have been bitten by a centipede, and generated discussions on such topics as the relative merits of beards worn by combat troops. The British commander-in-chief (Lord Raglan had died by this time and the post had been taken over by General Simpson), who had to reply to all the enquiries, is said to have remarked that 'The confounded telegraph has ruined everything'. The new French commander, Pelissier, who brought the war to a more or less successful conclusion, was not a friend of the telegraph either: 'It is impossible to carry on at the paralysing end of an electric cable' was his comment.

The next war in which the telegraph played a significant role was in India. The relative ease with which the British forces managed to suppress the Indian Mutiny was, to a considerable extent, due to their superior communications. For the first time in history military telegraphers were busy erecting lines following the advance of the commander-in-chief wherever he pitched his camp. John Lawrence, commissioner of the Punjab, is said to have exclaimed: 'The telegraph saved India.'

The first European war in which the telegraph moved with the troops was that between Austria and the French–Sardinian alliance in 1859. Napoleon III went with the troops this time. He needed the telegraph to send reports back to Paris and to keep in contact with his ally and with the other commanders. Since this particular war was of rather short duration, hardly more than six weeks, there was no time to build up a telegraphic corps. Those who built the lines were civilians.

I shall finish the story of the telegraph in wars with that of the American Civil War of 1861–5. The war started with specialist telegraphic corps already in existence, which were further expanded during the war. This was the debut of American technical know-how in war. They did it on a grand scale: 24,000 miles of lines were constructed, six and a half million telegrams were sent, and the whole thing cost over two and a half million dollars.

CHAPTER FIVE

The Telephone

Introduction

The mechanical telegraph was born in war. The electrical telegraph, although born in peace, soon became an instrument of war, and as has been seen, played a significant role in politics both internal and external. In contrast, the telephone made no major impact upon the battlefield. It might have made Mrs Jones an enemy of Mrs Smith on account of what she said about her to Mrs Robinson in the course of their lengthy telephone conversations but it did not in many other ways contribute to the spread of hostilities. Its effect on matters diplomatic was also slight. It rose to importance only in more recent times with the establishment of the hot line between Kruschev and Kennedy, between the Kremlin and the White House.

As a technical invention it was no doubt ingenious. There was a scientific background of course—all those working on the subject must have had a clear idea of the properties of sound—but still, drive and imagination turned out to be more important than scientific rigour. The contribution of science came only a quarter of a century after the establishment of telephone networks but then the contribution was crucial.

The story starts in Europe but those who brought the telephone to perfection and made it a commercial success all worked in America. Technical leadership, for the first time in the history of communications, passed across the Atlantic.

The Telephone Bursts upon the Scene in America

The man associated in the public mind with the telephone is Alexander Graham Bell and indeed he had a lot to do with the early introduction of the telephone. He became interested in problems related to speech early in life, stimulated by a long family tradition. His grandfather was a professor of elocution; his father, also a professor, wrote books on sound analysis for the purpose of teaching the deaf and dumb to speak.

He was born in Edinburgh in 1847 but moved to Canada with his family in 1870. He built his own laboratory in which he studied methodically the properties of sound. He was also interested in telegraphy, so it is quite natural that the combination of sound transmission with telegraphy led him to the telephone. By the age of 28 he had succeeded in producing a device that actually talked, and which he perfected and patented the

Getting the Message: A History of Communications. Second Edition. Laszlo Solymar, Oxford University Press (2021).
© Laszlo Solymar. DOI: 10.1093/oso/9780198863007.003.0005

next year.[1] It came to the attention of the public at the Philadelphia Exhibition in June 1876 where it was praised both by William Thomson and by Dom Pedro, the emperor of Brazil.

The demonstration which carried the fame of the telephone far and wide in the United States took place at the crowded Lyceum Hall in Salem, Massachusetts, on 12 February 1877. Bell had one of the telephones in the lecture hall, whereas the other one was in the hands of Watson, his assistant, 18 miles away in Boston. There was no need to lay a wire. In contrast to the telegraph, which had to be built from scratch, the telephone found a ready infrastructure. Bell and Watson could use the existing telegraph line between Boston and Salem. The demonstration started with Watson's rendering of 'Auld Lang Syne' and 'Yankee Doodle' followed by a conversation between Bell and Watson with others butting in. The demonstration was reported by several newspapers over the next few days. The scene was captured for posterity by *Scientific American* (see Fig. 5.1).

Commercial exploitation was not far behind. The man in charge of the operations was Gardiner Hubbard, Bell's future father-in-law. He first offered Bell's patent to Western Union, the company which had a practical monopoly in the telegraph business, for a sum of $100,000. The president of the company was not interested. He did not think the telephone would ever catch on (Fig. 5.2). So Hubbard had to start the business on his own. In May 1877, he announced proudly, to all and sundry:

The proprietors of the Telephone, the invention of Alexander Graham Bell, for which patents have been issued in the United States and Great Britain, are now prepared to furnish Telephones for the transmission of articulate speech through instruments not more than 20 miles apart. Conversation can be easily carried on after slight practice and with the occasional repetition of a word or sentence.

The terms for leasing two Telephones for social purposes connecting a dwelling house with any other building will be $20 a year, for business purposes $40 a year, payable semiannually in advance.

As it happened the telephone did catch on. It took Hubbard less than three months to acquire over a thousand subscribers. Not surprisingly, by the end of the year Western Union had a change of heart. They had the money; they got into the telephone business in a big way. Bell's company immediately issued a challenge. They claimed that Western Union had infringed their patent. The fight between David and Goliath took two years but at the end Goliath was duly smitten. The actual smiting was done by a judge who ruled against Western Union. A settlement between the contestants was reached in November 1879. Western Union agreed to withdraw from the telephone business for a period of 17 years and hand over all their telephones (over 50,000 by that

[1] As it happened he beat Elisha Gray, another telephone pioneer, to the patent office by two hours.

Fig. 5.1 Alexander Graham Bell demonstrating his telephone to an audience in Salem while maintaining contact with his laboratory in Boston.

time) and telephone patents (Gray's and Edison's among them) to their rival. In order to buy the equipment Bell's company needed new capital. The way to do it was to found a new company and issue new shares. After doing this several times, the one founded in 1885 was to stay. It was called the American Telephone and Telegraph Company, AT&T for short.

Manual exchanges were established as soon as the telephone was born. They spread quickly and gave job opportunities to lots of women. Had the growth gone on at the same rate for another half-century most of the United States' female population would have been employed as telephone operators. So some time in the twentieth century something

Fig. 5.2 A laughing stock? The president of Western Union did not think that the telephone would ever catch on.

had to be done about automation. Interestingly, the idea for an automatic exchange appeared as early as 1889, when one would still have expected a large number of young ladies to have been available. Indeed, the motivation came not from labour shortage nor from a desire to reduce the wages bill. It came because Almon Brown Strowger, an undertaker in Kansas City, found the number of orders for his wares declining. Since the rate of dying in Kansas City was fairly steady, Mr Strowger suspected foul play. And foul play it was. One of the telephone operators turned out to be the wife of the rival undertaker in the city, and she had been managing to divert Strowger's customers to her husband. Thinking of means to remove her from that key position Strowger invented the automatic exchange. It was a successful invention, first installed in 1892 and still used in many cities both in the US and in Europe as late as the middle of the 1970s.

When Bell's patent expired in 1893 other companies clambered upon the bandwagon but they could not seriously dent the monopoly of AT&T. It grew to become one of the biggest and most influential companies in the world. Of course, nothing lasts forever. AT&T lost its monopoly position in the 1980s but that's another story. It will be told in Chapter 13.

The Telephone in England, 1877–1911

The first Bell telephones were brought to Britain in 1877 by William Preece, who will appear in the narrative several times. He presented them to general public acclaim at the meeting of the British Association at Plymouth. He was quite keen to introduce a telephone service in the UK but he had some reservations. He did not think that Britons would take to the telephone as eagerly as the Americans because 'here we have a super abundance of messengers, errand boys, and things of that kind'.

It was relatively easy to make the first steps since the Post Office already leased a large number of private telegraph lines (connecting only two addresses), taken over at the time of nationalization from the Universal Private Telegraph Company. To use the telephone on the public lines would have been rather complicated: it would have constantly interfered with the telegraph traffic. But, of course, there was a case for erecting new lines specifically for the telephone.

While the Post Office and the Treasury were pondering about the best course to take, two companies, the Telephone Company and the Edison Telephone Company, were formed. They wasted no time. They started immediately to build lines and exchanges and advertised their services. Some of their methods earned the displeasure of the telegraph department of the Post Office. They complained that the companies were

straining every nerve to give to their business an appearance of success by carrying their wires to the premises of people who have never asked for them, and by offering to these people a year's use of their system free.

The department objected to these 'hard sell' techniques because they were highly suspicious of the companies' motives. They believed that the main aim was to invest vast sums in building a telephone network and then to compel the Post Office to buy it at inflated prices.

So did the Post Office decide to build their own lines and exchanges and sell their services so cheaply as to bankrupt the companies? No, they did not choose that option. They were afraid of the risks. By this time the telegraph department had hardly made any profit at all (it had, in fact, made heavy losses if the servicing of debts, incurred at nationalization, was taken into account). They were not keen to incur possible further losses by the introduction of the telephone.

Did, perchance, the Post Office decide to do nothing? No, they did not take that option either. They could not possibly give a free hand to the companies because, to use a modern phrase, the telephone and the telegraph were substitutes for each other. A message sent by telephone was one message less for the telegraph. The advance of the telephone companies clearly threatened the business of the telegraph department. So they had to do something. Slowly, they started to build some lines, mainly trunk lines between cities and exchanges. The main thrust of their strategy was, however, a legal challenge to the companies. They claimed that by setting up a telephone network the companies violated the Post Office's telegraph monopoly. The companies believed they did no such thing. They pointed out that the telephone was a new invention which had nothing to do with the telegraph.

The Telegraph Act of 1869 referred specifically to the monopoly of sending telegrams. Does a message sent by telephone constitute a 'telegram'? That was the question. The legal moves took quite some time. It was only in December 1880 that judgment was delivered. The judges found in favour of the Post Office.[2] The companies (several new companies were founded in the meantime, and the original two amalgamated to form the United Telegraph Company (UTC)) reluctantly admitted defeat and expressed willingness to come to some compromise with the Post Office.

Agreement was reached in April 1881. The terms of the Post Office were quite hard. They reserved for themselves the lucrative business of building trunk lines. To the UTC they offered operation within a radius of 5 miles from the centre of London for a royalty of 10 per cent of the gross receipts, and the licence was to run for 31 years, subject to a possible termination of the agreement on the 10th, 17th, or 24th year should the department so wish.

The companies accepted the terms but hostilities did not stop. They degenerated into guerilla warfare. The Post Office kept on putting various restrictions upon the companies. They did not want the companies to flourish but they did not want them to perish either (after all they paid a 10 per cent royalty). The story of the skirmishes was described by Hugo Meyer.[3] According to a reviewer of the book in the *Electrician*

The reader who goes carefully through this book will get a vivid impression of the evil results of State monopoly of a trading industry, and can hardly fail to lose a portion of what respect he may have had for the intelligence, foresight and commercial acumen of British politicians.

The companies soldiered on. Their licenses did not even include the right to erect poles in the streets or to lay cables underground. The wires had to be strung from house-top to house-top. They were not a pretty sight. The Duke of Marlborough gave voice to these concerns in July 1889 in the House of Lords:

There was no single town in civilised Europe which was so behindhand in its telephone communication and none which had been so obviously spoilt and disfigured by wires running in all directions than London.

The sad state of the telephone service earned the displeasure of *Punch* too. One of their cartoons in 1892 depicted the telephone as Cinderella, whereas postal and telegraphic services appeared as the Ugly Sisters (see Fig. 5.3).

The change over from house-tops to streets was made possible by an Act of Parliament in 1892 but, owing to the objection of the local authorities, it was not put in practice for a number of years afterwards. Unfortunately, when permission was granted, a considerable proportion

[2] The ambiguity was removed this time. All transmission of signals, whether wires were used or not, constituted a violation of the Post Office's monopoly.
[3] *Public Ownership and the Telephone in Great Britain* (MacMillan, New York, 1907).

PUNCH, OR THE LONDON CHARIVARI.—April 2, 1892.

THE TELEPHONE CINDERELLA;

OR, WANTED A GODMOTHER.

Fig. 5.3

of the lines built were the overhead types which had no aesthetic appeal either. The photograph shown in Fig. 5.4a was taken in 1910. Their encroachment on property and skyline induced *Punch* again (Fig. 5.4b) to register a protest.

It might be worth digressing here a little and note that the French were less willing to spoil their capital city. The cables were put in the bowels of the Earth alongside the sewage system (Fig. 5.5). The Americans, on the other hand, fared even worse. Broadway in 1890 was not a pretty sight (Fig. 5.6).

Returning now to England and the tug-of-war between the telephone companies and the Post Office: it ended with the complete victory of the latter. In 1901 an agreement was reached allowing the Post Office to purchase the assets of the companies (owing to further amalgamations this was by then the National Telephone Co.) at valuation and without goodwill when the licence expired in 1911. The eventual cost to the Post Office was £12.5 million in contrast to the £20 million demanded by the

(a)

Fig. 5.4 (a) Telephone poles in Besses o' th' Barn, 1910.

company. The Post Office could claim that this time they were not taken to the cleaners. Public money was saved but the telephone network was in shambles.

Women's Employment

The manual exchanges needed operators. At the beginning, both in the US and in Britain, some attempts were made to employ young boys but the experiment did not last long. Boys turned out to be loud, undisciplined, unreliable, and rude to customers. Very soon, and in practically every

Fig. 5.4 (b)

(b)

country, exchanges were operated by women. The female voice was supposed to be clearer; the women were more patient, more forbearing, more attentive to the customer's needs, and, of course, cost a lot less than men whose employment was mostly confined to night duties. Women were also believed to be more docile, and less likely to join a trade union. The working conditions were quite hard to begin with. The operators had to move around to reach the switches (Fig. 5.7a). However, later designs allowed the women to do their work sitting (Fig. 5.7b).

Marriage was frowned upon. Girls were expected to leave when they married. Unfortunately, the very qualities which made them good switchboard operators—docility, patience, adaptability, lack of ambition—also made them good candidates for men's attentions. As a result turnover was rather high, far too high for the liking of the companies. A remedy that most companies adopted was to provide various perks ranging from medical attention to comfortable restrooms.

(a)

Fig. 5.5 The telephone cables
went underground in Paris.

(b)

Fig. 5.6 Broadway and John Street, New York City in 1890.

In Germany, Princess Louise, daughter of Kaiser Wilhelm II, campaigned for women's rights in general and for their right to work in telegraph offices and in telephone exchanges in particular. She won, but not without a fight. As a result of her victory the switchboards were slowly taken over by women after 1890. There were, however, some regulations concerning dresses. They had to reach 20 cm below the knees. A view on how the inspection was carried out is shown in Fig. 5.8.

The number of telephone operators rose fast all over the world. By the 1920s giant exchanges employing over a hundred women were not uncommon. London's Gerrard exchange in 1926 is shown in Fig. 5.9. That was about the time automatic exchanges started to multiply and the long decline of the number of women operators began. At the time when the first edition of this book was written (late 1990s), there were still a few women in their original jobs connecting person to person, but the great majority have moved on to other parts of the business. When you make a complaint about any kind of deficiency in the telephone service the likely thing you will hear is the gentle voice of a woman endowed with patience and forbearance. Docility is rather rare nowadays.

(a)

(b)

Fig. 5.7 Early telephone exchanges in Paris: (a) operators work standing, and (b) operators are seated but still need to stand up for reaching the upper parts of the switchboard.

International Comparisons

Since the telephone was first established in the US it is not particularly surprising that they kept that lead afterwards. What is more surprising is that Europe followed suit rather slowly, hesitantly; one might even be tempted to say, reluctantly. There were of course reasons for this. The countries of Europe were smaller; hence, postal services were faster and

Fig. 5.8 An inspector checks that a German operator's dress is at least 20 cm below the knee.

perhaps Europe was in less of a hurry. A few days here or there did not make such a difference, and if something was really urgent there was always the telegraph which had the advantage of providing written evidence. All these factors must have played a role in the slow start and slow growth which was characteristic not only in the United Kingdom but in other industrially developed countries as well.

The growth in the number of telephone subscribers between 1885 and the beginning of the First World War is shown in Fig. 5.10. It is plotted on a logarithmic scale which makes it possible to show the rapid growth. It may be seen that by 1914 the number of subscribers in the US reached 1 in 100 of the population. As may be seen, the US always had the highest number of telephones, except for a short interval when the lead was gained by Sweden. The reason for the slow growth in America was the lack of competition. But when Bell's patent expired, a number of new companies entered the fray, resulting in much faster growth. By 1900 the US regained the lead.

Fig 5.9 The Gerrard
exchange in London in 1926.

Why was Sweden the most telephone-minded European country? Firstly because it was highly industrialized, secondly because the state interfered much less than in other European countries, and thirdly, perhaps, because of the long winter nights. The telephone might have been the best means to relieve solitude. Denmark, another Scandinavian country, with somewhat brighter winter nights and with an equally non-interventionist government, was in second place among European countries.

The performances of Germany and of the UK were fairly similar. The very small growth in the UK between the years 1905 and 1911 may be attributed to impending nationalization. The figures for France were remarkably low. This was partly due to the dead weight of the state but that couldn't have been the full reason. Maybe their promotional campaign was poor. They advertised the entertainment more than the business aspects and presumably the French preferred other kinds of entertainment. Austria and Italy lagged even further behind, which may be explained by their lower levels of industrialization.

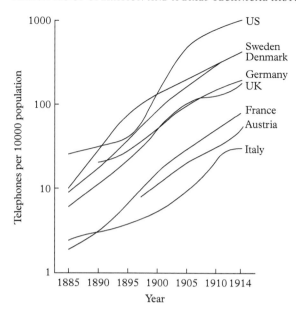

Fig. 5.10 An international comparison of the growth of telephone subscribers in the period 1885 to 1914.

Principles of Operation and a Brief Technical History

Are the basic principles as simple as those needed to understand the telegraph? Not quite. To understand the telegraph all one needs to know is that electrical circuits may be switched ON and OFF and, in addition, one must be able to distinguish the symbols 1 and 0 from each other.

In order to understand the basic principles underlying the operation of the telephone somewhat more is needed. One should, to start with, have some idea of the physical properties of sound. What are they? By the middle of the nineteenth century everything that is important was known about sound and the way we hear. Some of it was fairly obvious (that, for example, the mouth works as the transmitter and the ears as receivers) but the more detailed mechanism was also known. Sound is carried by vibrations in the air which are conveyed by the ear drum to the nerves by a series of bones called the hammer, the anvil, and the stirrup. Conversely, when we utter a sound the air is brought into vibration. As the vibrations propagate they become gradually weaker and weaker as they spread further from the source. Consequently, if the ear is too far away from the source of the sound, then the vibrations are not strong enough to give the sensation of hearing. How could this limitation be overcome? There are several possible solutions. One is to use the good offices of electricity, and that is the one this chapter is mainly concerned with. There are, though, other possibilities too, some of them going back to ancient times.

(a)

(b)

According to the researches of Athanasius Kircher in the Vatican Library,[4] Alexander the Great called together his army, scattered over a large area, by an enormous double-ringed horn (see Fig. 5.11a). Kircher also realized that sound can be directed with the aid of ducts in any direction, and a ruler interested in the idle gossip of his subjects might want to make use of such a system in his palace, as shown in Fig. 5.11b.

Although ducts can be used for guiding sound, their disadvantage is that the sound still travels in air in which it quickly fades away. An alternative solution is to let the sound propagate in a medium in which it does not decline so fast. A taut wire would serve for the purpose. In fact, Robert Hooke, who appeared in Chapter 3 as one of the pioneers of the mechanical telegraph, made experiments with sound as well. In 1667 he reported:

I have by the help of a distended wire, propagated the sound to a very considerable distance, in an instant, or with as seemingly quick a motion as that of light, at least incomparably quicker than that which at the same time was propagated through the air; and this not only in a straight line or direct, but in one bended in many angles.

Similar experiments using metal rods were conducted in the 1820s by Charles Wheatstone, the man who played such a crucial role in the birth of telegraphy. He was able to communicate this way between chambers. He thought the problem lay in finding the right conducting substance in which the sound declines even less than in metal rods. If that substance was found, he noted optimistically, 'it would be as easy to transmit sounds from Aberdeen to London as it is now to establish communication from one chamber to another'. Unfortunately, those substances have not been found so far.

The conclusion is that, although the direct transmission of sound appears to be possible, it does not seem to be a practical solution.[5] So it was a natural thought to turn to electricity for help in the nineteenth century. In order to see what the first inventors did, some relevant properties of sound need to be discussed.

The two most important properties are intensity and pitch. Intensity is an easy concept to grasp. When we speak loudly the intensity is high, when we speak quietly the intensity is low. What about pitch? It is related to the number of vibrations per second. Vibrations of what? Of the atoms in a metal rod, for example. When sound propagates in a metal rod then each of the atoms of the rod moves around its rest position as shown schematically in Fig. 5.12a. This motion relative to the rest position is plotted in Fig. 5.12b as a function of time. The atom moves away in one direction, reaches its extreme value after which its motion is reversed and it moves back to its rest position which it overshoots, reaches its extreme in the other direction reverses its motion again, etc., etc. This curve is described as sinusoidal. The maximum deviation of

Fig. 5.11 (a) A double-ringed horn, used allegedly by Alexander the Great. (b) Eavesdropping with the aid of ducts.

[4] Taken from his book *Phonurgia nova* published in 1673.
[5] It might have been a practical solution at certain times in certain societies. According to a report in *The Illustrated London News* (30 July 1910) entitled 'Marconigrams of the Congo: Batetela Gong-Signalling', their correspondent came across a communications system capable of transmitting any message. The following description is given: 'The gong used by the Batetela for sending messages is first cut from one large solid piece of hard wood. It is then hollowed out, the whole of the interior being removed through the long opening at the top. The hollow inside follows the outer shape. The sticks used to beat the gong have at their end a knob of rubber. To send a message, the beater of the gong will ascend a hill in the evening. The sound of the drum, very rough when near-by, is quite beautiful music at a distance. I have tried the abilities of these drummers by having a message drummed to a village six miles distant asking the chief to 'send me the arrow he showed me the evening before: not the one with an iron tip, but the one with the twisted feathers.' The arrow arrived in less than an hour. This gong, a solid block of wood, gives three sounds on each side, according to where it is beaten. The six sounds so obtained are used to form a syllabic alphabet which permits them to transmit messages, however complicated they be. The sound carries about seven miles.'

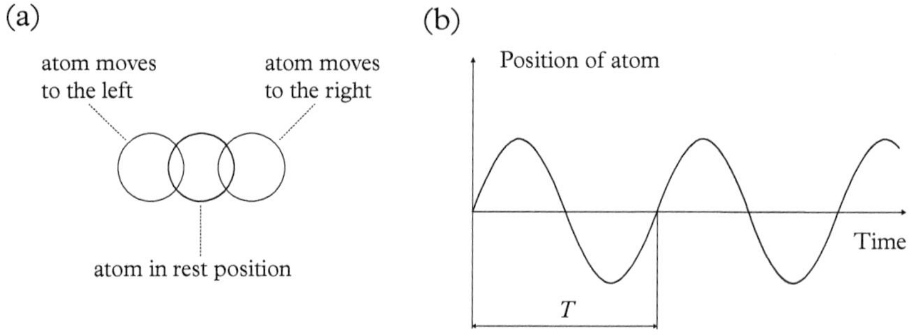

(a)   (b)

atom moves   atom moves   Position of atom
to the left   to the right

atom in rest position   Time

T

Fig. 5.12 (a) Atoms vibrating. (b) The position of an atom relative to its rest position. T is the period of vibration.

the atom from its rest position is called the amplitude. It may further be noticed that the curve reproduces itself again and again.[6] To execute one cycle takes a time T, which is called the period. The pitch is given by the number of periods in a second. If the period is 1/100th of a second, then there are exactly 100 periods in a second. One may say then that the pitch is one hundred cycles per second. In technical language it is more usual to say that the frequency is 100 Hz.[7] The human ear is sensitive to sounds within the range of about 20 to 15,000 Hz.

Let us imagine now that we are familiar with the telegraph and our aim is to invent the telephone, relying again on electric current. We know that in the case of the telegraph, information is transmitted by breaking the current in an electric circuit. If the main information about sound is contained in its frequency, would it not be possible to transmit sound of a certain frequency by switching the current on and off periodically at the same rate? The first man who thought of it was Bourseul. In a paper, published in 1854 in *L'Illustration*, he states:

I have asked myself whether speech itself may not be transmitted by electricity—in a word, if what is spoken in Vienna may not be heard in Paris. The thing is practicable in this way...Suppose that a man speaks near a movable disc sufficiently flexible to lose none of the vibrations of the voice; that this disc alternately makes and breaks the current from a battery; you may have at a distance another disc which will simultaneously execute the same vibrations.

Bourseul refers to some experiments in his paper but had they been successful surely someone, somewhere would have mentioned them. The honour of producing the first telephone goes to Philipp Reis, a teacher in Friedrichsdorf who is unlikely to have ever come across Bourseul's work. He presented a paper before the Physical Society of Frankfurt am Main in 1861. A more detailed version of the instrument (shown schematically in Fig. 5.13) was described in 1862 by Legat, an inspector of the Royal Prussian Telegraphs. There is a membrane which is brought into vibration by the incident sound. A strip of metal cemented to the apex of the membrane is part of an electrical circuit.

[6] In mathematical jargon it is called a periodic function.
[7] Hz is the abbreviation for hertz after Heinrich Hertz, a German scientist whom we shall meet later when coming to wireless telegraphy.

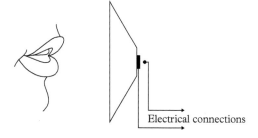

Fig. 5.13 The principles of Reis's telephone. The diaphragm vibrates in response to the incident sound making and breaking the electrical circuit.

Electrical connections

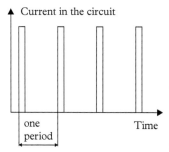

Current in the circuit

one period
Time

Fig. 5.14 When contact is made, an electric current flows.

When the membrane moves forward a sufficient distance it touches the contact point and the circuit is closed.

There is no doubt that Reis's telephone was a working device. It was even shown to the emperor of Austria and to the king of Bavaria in 1863. Unfortunately, it was not a device for which people would have been willing to open their purses. The quality was poor. It was apparently good for reproducing melodies but had some difficulties with the vowels in human sound. Many copies were produced at the time and sent all over the world. Bell might have seen one in his early youth; a copy is known to have reached Edinburgh, but he always denied any knowledge of it. Some of these telephones survived and can be seen in museums. There is one in the Science Museum in London.

If this telephone operates by switching the current on and off, what will the current be like when a pure sound at a certain frequency is to be transmitted? If the frequency is 100 Hz, then the current will be switched on once in every 1/100th of a second. The corresponding current as a function of time is shown in Fig. 5.14. The trouble is that this does not look like the sinusoidal curve in Fig. 5.12b at all. So it is not surprising that the Reis telephone did not work very well. It clearly needed some improvement. A clue as to how this improvement could be achieved is provided by Ohm's Law, stating that for a given voltage the current is inversely proportional to the resistance. High resistance means low current and vice versa. So if the resistance varies in a sinusoidal manner (Fig. 5.15a), the current will follow suit as shown in Fig. 5.15b (note that the current is maximum where the resistance is minimum).

The problem has now been reduced to finding a device in which the resistance varies sinusoidally in response to a single tone of sound incident upon it. A possible solution is shown schematically in Fig. 5.16. The front carbon and the back carbon are made of solid material. They are good conductors. The carbon granules between them are compressible. As they are compressed their resistance declines and the current increases in the electrical circuit. Thus, the variation of incident sound pressure will cause corresponding current variation which will then be

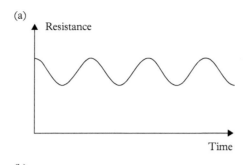

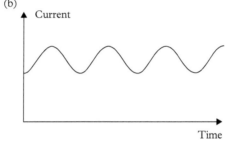

Fig. 5.15 (a) The electrical resistance varies sinusoidally in response to the incident sound. (b) The current follows suit.

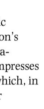

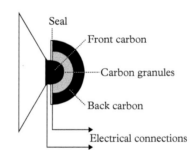

Fig. 5.16 A schematic representation of Edison's microphone. As the diaphragm vibrates it compresses the carbon granules which, in response, change their resistance.

available at the other end of the line. How can it be converted back into sound? The device that will do this has already been introduced in Chapter 4. It is an electromagnet that attracts a piece of iron. The difference is now that the piece of iron has been replaced by a thin iron diaphragm. If the current in the coil varies sinusoidally, the diaphragm moves sinusoidally, forcing the air into motion which is perceived as sound. When the input sound is not a pure sinusoidal wave but represents human speech it is still faithfully reproduced.

That is about it. The receiver described above was invented by Bell and the transmitter by Edison (it became known as the microphone). After a few years (that is from the 1880s) they became the standard components in practically every telephone apparatus and, in fact, they were in use until very recent times.

The CR Limit and Periodic Loading

Sound declines as it moves away from its source. That much is obvious. What may be added here is that when the electric current is modulated by the incident sound, that modulation will decline too. At the transmitting end the current as a function of time looks like that shown in Fig. 5.17a, but at the receiving end the modulation may be much smaller, as may be seen in Fig. 5.17b. The modulation may, in fact, be so small that it would be difficult to distinguish it from noise. How long can the line be before the information is lost? It may be worth remembering that in Hubbard's announcement, telephone lines were offered for distances not exceeding 20 miles. So 20 miles, using overhead wires, was OK.

In engineering jargon the effects responsible for limiting the distance are attenuation and dispersion. Attenuation is just another word for decline. Dispersion is, however, a more sophisticated phenomenon. When we talk, and particularly when music is played, some sounds have higher frequencies than others. The phenomenon that different frequencies may travel at different speeds is called dispersion. If some frequencies arrive before the others, that will make speech unintelligible and music unenjoyable. This problem of telephony turned out to be very similar to that of telegraphy analysed by William Thomson in 1855. According to his calculations the distance that can be achieved in telegraphy depends on the product of resistance (R) and capacitance (C), which must be below a certain limit. When it came to telephony this CR product was found to play a similar limiting role. Thus, in order to increase the effective distance of telephone lines, either the resistance or the capacitance, or both, had to be

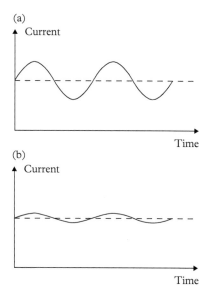

Fig. 5.17 (a) The modulated electric current. (b) At the other end of the wire the modulation is smaller.

reduced. One could reduce the capacitance by using overhead lines instead of cables. That much was well known. The resistance could be reduced simply by increasing the diameter of the wires. This was tried and it did indeed help, as witnessed for example by the successful operation of the New York–Chicago and Paris–Marseille lines. There was, however, an economic limit to using thicker and thicker wires. It was too much of a burden, both on the finances and on the supporting poles. The limit with overhead lines was believed to be about 800 miles. This was a very large distance but still not enough for lines to stretch from the Atlantic to the Pacific. For underground cables the limit was considerably less, about 30 miles.

Whenever a limit is reached it is the engineer's job to find a way round it. William Preece, the chief engineer of the British Post Office from 1877 to 1899, strongly believed in the inviolability of the CR limit. After all, both the calculations of William Thomson and practical experience pointed in the same direction: a limit clearly existed. Oliver Heaviside, a self-made theoretician, thought otherwise. He recognized that Thomson's 1855 theory of cables, as it stood, was not applicable to the telephone. For the telegraph it was indeed sufficient to take into account resistance and capacitance, but for the telephone the correct model had to include a third parameter, the so-called self-inductance (or simply inductance) as well. Heaviside agreed that the limit did apply to the existing telephone lines but, according to his theory, it could be significantly extended by inserting inductances (coils wound on an iron core) into the lines at periodic intervals. He wanted to announce this discovery in a joint paper with his brother who was a telegraph engineer fairly high up in the hierarchy. Preece, occupying the highest technical position at the Post Office, refused permission to publish the paper.

A major clash between proponents of the old and new theory came at the 1888 Bath meeting of the British Association on the seemingly unrelated subject of lightning conductors. The man who entered the ring as an ally of Heaviside was Oliver Lodge, a professor of physics at Liverpool University. According to Lodge, self-induction had to be introduced if one wanted to describe all the properties of lightning conductors. Preece was unable to accept that self-induction played any role and he expressed his doubts in no uncertain manner:

The practical man, with his eye and his mind trained by the stern realities of daily experience, on a scale vast compared with that of the little world of laboratory, revolts from such wild hypotheses, such unnecessary and inconceivable conceptions, such a travesty of the beautiful simplicity of nature.

The fierce debate between Preece and Lodge at the Bath meeting ended, according to *Electrical Plant*, with the victory of Preece, who managed to slay his opponent (see cartoon of Fig. 5.18).

Fig. 5.18 Caricature of W. H. Preece (standing) and Oliver Lodge.

As it turned out later, Heaviside and Lodge were right and Preece was wrong. Was Preece's predicament the same as Whitehouse's in the 1850s? It may be worth remembering that Whitehouse called Thomson's calculations on the operation of the submarine cable a 'fiction of the schools'. Whitehouse just could not imagine that sophisticated mathematics had predictive power. Thirty years later, the situation was quite different. Thomson's calculations had by then been incorporated into the conventional wisdom. Mathematics was no longer a dirty word. In fact, Preece boasted that he had made 'mathematics his slave' in contrast to his opponents who were 'the slaves of mathematics'. Preece had no objection to science either. He is actually on record as having said that 'the engineer must be a scientific man'. So why was he in the wrong camp?

There could have been three reasons. Vanity must have played a significant part. He was not to be told by people outside the Post Office how to run his department. The second reason was that the part of mathematics that Preece made his slave was very respectable for an engineer at the time but, unfortunately, it was hopelessly inadequate for understanding Heaviside's theory. In the third place, Preece did not understand the physical reasons for the significance of self-induction. For that he would have had to study Maxwell's theory, of which he was entirely ignorant. If one wanted to be kind to Preece one could say that he had an inordinate amount of bad luck. He was caught up in the whirlwind of the greatest advance in the history of science since Newton produced his theories on the motion of heavenly bodies. That great advance was due to Maxwell, but Heaviside and Lodge played a crucial role in making Maxwell's electromagnetic theory acceptable to the rest of the world. I shall return to Maxwell in Chapter 6 on wireless telegraphy where the discussion properly belongs.

In Britain the discussions on self-inductance were strictly academic. There was no pressure on the Post Office to find ways for extending the limit of telephones. The existing lines worked satisfactorily. The situation, as mentioned previously, was quite different in the US. There was plenty of motivation there to try any new method for extending the useful range of the telephone. The American Telephone and Telegraph Company was well aware of the likely financial rewards but they did not as yet have the people who could address the problem from scratch. They did, however, have men capable of understanding, digesting, and developing further existing theories. In contrast to Preece, they recognized the significance of Heaviside's calculations and started to test his ideas experimentally. It was a long slog. From conception to success the work took about ten years. It was only around the turn of the century that the engineers of AT&T convincingly demonstrated the beneficial effects of the periodic insertion of coils, a technique that became known as periodic loading.

AT&T filed a patent in March 1900. It described the way the coils had to be inserted in the lines but stopped short of giving a mathematical expression for how large the coils should be and at what intervals they should be inserted. Two months later Michael Idvorsky Pupin filed a similar patent in which all the mathematical calculations were given.

Pupin was a Serbian immigrant, one of the huddled masses of Europe who decided to seek their fortune in the US. He arrived in New York, penniless, at the age of 15. Working while studying, he became a student and later a professor at the prestigious Columbia University. He had plenty of paper so he could derive mathematical expressions, but he did not have the resources to do any experiments.

The patent situation was not clear at all. AT&T's was supported by experimental results but Pupin's gave all the information needed. AT&T took legal action, but without waiting for the outcome they decided at the end of 1900 to buy Pupin's patent. The significance of the invention may be appreciated by the sum AT&T was willing to pay. Pupin received $185,000 immediately, plus a further $15,000 per year during the 17-year life of the patent. The lessons of the story were not lost on AT&T's management. It was probably the last time that they were beaten by a lone scientist.

The advantages gained by periodic loading were quite significant. It made possible the opening of the 4,300-km line from New York to Denver in 1911 but that was about its limit. The transcontinental line (opened in 1915) had to wait for the invention of the electronic triode amplifier. For more about electronics and amplification see Chapter 6 on wireless telegraphy.

Digital versus Analogue

The terms digital and analogue are no longer reserved for the use of communications experts. They are freely bandied around in everyday conversation. It is easy to explain what they mean. Digital information is in the form of digits, e.g. 5 or 8 or 3.5 or 7.969. Some information may only exist in digital form, e.g. the number of houses in a street, the number of bedrooms in a house, or the number of jokes told at a party. Examples of analogue information are length, weight, the time on an old-fashioned watch, the speed of a vehicle or the current flowing in a circuit. All these, at least in principle, could be measured with arbitrary accuracy. We could, for example, measure the distance between two points as 12,365 m. A more accurate measurement for the same distance might yield 12.36513 m. The point is that the distance is expressible with digits but there is no exact digital equivalence. The higher the required accuracy, the greater the number of digits needed.

It follows from the definitions so far that telegraphy uses digital signals, whereas telephony belongs to the analogue family. It has to reproduce the infinite variations of a sound pattern characteristic of a particular individual. Examples of digital signals have been given before. The variation of current describing the word 'eat' in Morse code (dot, dot-dash, dash) is shown in Fig. 5.19a, which includes spaces between the symbols. I can just as well show here (Fig. 5.19b) another possible representation when dots and dashes are coded by currents in different directions. This latter code may be seen to be more efficient. It sends the same message in less time.

In contrast, sound is an analogue signal. The intensity of sound varies continuously as a function of time. It might take the form shown in Fig. 5.20. It is a rather complicated signal so it is not surprising that the telephone of Philip Reis did not work very well. There is certainly more to telephony than the breaking and making of current in an electrical circuit.

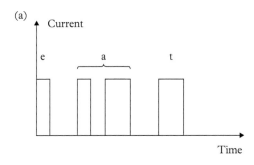

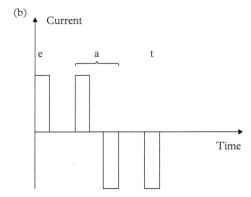

Fig. 5.19 The word 'eat' as rendered in Morse Code by (a) dots and dashes, (b) by opposing currents.

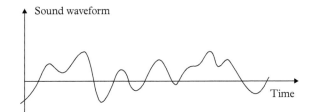

Fig. 5.20 The amplitude of sound as a function of time.

Having come to the conclusion that analogue and digital signals are quite different, it might still be worth asking the question: 'Is there some relationship between them? Can they be converted into each other?' Nineteenth-century electricians would have said, no. Modern communications engineers say, yes. In the latest communications systems analogue signals are converted to digital ones at one end, then transmitted as digital signals, and finally they are reconverted to analogue signals at the other end. Is that a good thing? It is, because the achievable sound quality is much better. I shall return to this problem in Part III.

Telephone Broadcasting

Broadcasting, as the name suggests, means sending information from a central point to an audience far and wide.[8] It became popular with the

[8] I said in the Introduction that I had no intention of including broadcasting in this book because it covers too wide a field. I shall make an exception here with telephone broadcasting partly because it is a relatively unknown episode in the history of communications, but perhaps mainly because it gives me a chance to include my native city in the narrative.

advent of radio but there was never any reason why the telephone should not have played the same role by laying wires to every subscriber. As early as August 1876 an article in *Nature* predicted that

by paying a subscription to an enterprising individual who will, no doubt, come forward to work this vein, we can have from him a waltz, a quadrille, or a galop just as we desire.

Several attempts were made in Britain and in the US to provide such entertainment but none of them had a great number of subscribers, and none of them remained solvent for long. There was, however, one exception: the *Telefon Hirmondo* (Telephone News, in Hungarian), which was founded in Budapest in 1893. It was the brainchild of Tivadar Puskas, who became well aware of all the modern techniques by working for a while in Edison's laboratory in America and later becoming his representative in Europe.

The enterprise started as a 'news only' service with the news repeated at the beginning of each hour, but by 1896 it had a full programme:

9.30–10.00	Programme parade. News from Vienna and from abroad. Announcements of the official gazette.
10.00–10.30	Report from the stock exchange
10.30–11.00	Review of the papers.
11.00–11.15	Report from the stock exchange
11.15–11.30	Local and theatre news. Sport.
11.30–11.45	Report from the stock exchange
11.45–12.00	Parliament, news from the provinces and from abroad
12.00–12.30	Parliament, court, political and military news
12.30–1.30	Report from the stock exchange
1.30–2.00	Repeat of the more interesting news
2.00–2.30	Parliament, local government news, telegrams
2.30–3.00	Parliament, local news, telegrams
3.00–3.15	Report from the stock exchange
3.15–3.30	Feature articles from the papers
3.30–4.00	Parliament, weather, news from the courts
4.00–4.30	Report from the stock exchange
4.30–5.00	Review of Viennese papers, financial and economic news
5.00–6.00	Children's programme on Thursday. On other days
5.00–5.30	Theatre, sport, art, fashion, literature
5.30–6.00	English lessons on Monday, Wednesday, and Friday
6.00–6.30	French lessons on Monday, Wednesday, and Friday
6.00–6.30	Italian lessons on Tuesday, Thursday, and Saturday
7.00–	Direct broadcasting either from the Opera House or from the Folkstheatre. In the interval after the first act, reports from the Paris, Berlin, and Frankfurt Stock Exchanges

After the end of the performance general news and stock exchange reports were broadcast, followed by military and gipsy music until midnight. On Sunday afternoon there was a grand concert directly from the studio.

Among the services provided was a time signal before each hour which lasted for exactly 15 seconds. There were a few advertisements too, in the breaks between programmes. For example, on 19 February 1893 it was announced: 'Guaranteed teeth, 1 forint 50 a piece. Dental Laboratory, Apfel, Budapest, Erzsebet ter 10.'

Was the popularity of the *Telefon Hirmondo* enhanced by any association with murder, as happened earlier with the telegraph, and as will happen again when the availability of radio contacts will be responsible for the arrest of a murderer? I could find reference only to a rather minor crime, committed by a Mr Kapus, in one of the contemporary newspapers. In fact, it was not clear at the time whether it was a crime at all. This is the usual case of technical progress bringing new legal problems in its wake. An earlier example, the activity of the Blanc brothers, has already been mentioned. They used the official Paris–Bordeaux mechanical telegraph line for their own profit. They were acquitted in the end because the prosecution could not find the right paragraph of the law. The alleged crime of Mr Kapus was in a different category. He connected a wire to the *Telefon Hirmondo* line passing under his window and listened illicitly to the programme. The legal problems were connected partly with property rights (who owns the space under the window) and partly whether sound can be stolen. According to the newspaper report the trial was adjourned because the judge wanted further legal advice. I would be only too happy to satisfy the reader's curiosity as to the fate of Mr Kapus but, I regret, I could not find any further legal reports on the case.

What was the technical quality of the broadcast? In the beginning there were quite serious problems with clear sound transmission, but great improvement came a few years later with a specially designed system which bypassed the regular telephone network. It was built according to the original designs of Puskas who died very soon after the first broadcasts. He had an ingenious technical solution which ensured high sound quality and made the network economic to install. Each district had its own substation which obviated the need to lay a wire from the studio to each subscriber separately.

The *Telefon Hirmondo* employed about 150 people and by 1896 it had about 6,000 subscribers, a number which fluctuated a little (reaching a peak of 7,629 in 1899) but remained fairly steady until the beginning of the First World War. It reached its low point in 1915 with 2,821 subscribers, from whence it picked up again. During the revolutions of 1918, the following chaos, and the occupation of Budapest by Romanian troops, the wires were mostly destroyed. Broadcasting restarted in 1924, and a year later the *Telefon Hlrmondo* generously allowed the new upstart, the

Radio Hirmondo, to use their studio. Interestingly, the two media were never in competition. After a while they agreed to broadcast the same programme. The number of subscribers who preferred to listen to the telephone, rather than to the radio, did actually increase to over 10,000 during the 1920s and 1930s. The final demise of the *Telefon Hirmondo* came in 1944. It was not resurrected after the Second World War.

Why was telephone broadcasting a success in Budapest and nowhere else? First of all its success was due to its inventor, Tivadar Puskas. He saw the need, offered an elegant technical solution, and raised the necessary capital. Who were the subscribers? Those who were interested in prices of stocks and shares were obviously among them; otherwise, there would not have been so many reports from the stock exchange.

Fig. 5.21 A poster advertising the *Telefon Hirmondo* in 1907. 'The Hungarian Royal Opera and the Folkstheatre-Comic Opera may be heard alternately on all stations of the Telefon Hirmondo. May be ordered at Rakoczi street 22. Installation free, subscription for a year 36 Korona.'

Who else? The prime minister and every member of his cabinet; the mayor of Budapest; Mor Jokai, the doyen of Hungarian writers; and everybody who was anybody. The *Telefon Hirmondo* also became a permanent feature in coffeehouses, hospitals and hotels, lawyer's offices, in the waiting rooms of doctors and dentists, and even in the better barber shops.

So what were the reasons for its success? They were partly economic. It was a time of great economic progress. Budapest could boast both of an urbanized aristocracy and of a prosperous middle class. There was no shortage of people who could afford the subscription fee. The cultural reasons were perhaps just as important. The same people who could pay the subscription were also interested in listening to the opera. And indeed in the publicity material (see Fig. 5.21) the opera was strongly emphasized. And finally, leisure time. The rat race had not been invented yet—not in Hungary, anyway. The middle class had to work—but not too much. They had plenty of time for entertainment.

The nearest thing to the *Telefon Hirmondo* elsewhere in the world was the Electrophone Company founded in England in 1894. It provided only entertainment, plays from the more prestigious theatres, and sermons from the more prestigious pulpits. It never went as far as to provide news and it never had more than 600 subscribers. Why not? Was it the Protestant ethic which did not allow successful businessmen to enjoy the fruits of their labour? Or was it just the opposite? There was so much entertainment available in London that they did not need to rely on the Electrophone Company!

CHAPTER SIX

Wireless Telegraphy

Introduction

There are a number of major milestones in the history of communications and, in our narrative, each of them has so far been associated with a single country: the mechanical telegraph with France, the electrical telegraph with England, and the telephone with the United States. There were, of course, good reasons for this. Revolutionary France, facing a hostile Europe, was ready to explore any new means likely to help the war effort. In England in the 1830s the newly established railways cried out for fast communications in order to avoid accidents and to improve efficiency. In response, William Fothergill Cooke and Charles Wheatstone managed to turn an academic toy into a practical device. What about the telephone? Shouldn't it also have started in England, the leading industrial power at the time? Not necessarily. In England the needs of industry and business were perfectly well served by the Royal Mail and the telegraph. And perhaps temperament also played a role. To make instant decisions, without the time to reflect, was not an Englishman's way of conducting his affairs. The telephone's success needed a country with vast open spaces, vast distances, and a population less inclined to hide their feelings.

The next great invention in communications was that of the wireless. In which country should the inventor seek his fortune? First, perhaps in his native country, but if that one is not very industrialized what should his next choice be? Would it be the US where opportunities were the greatest? No, opportunities themselves are not enough. He would try to sell his invention in the country which had the greatest number of ships, and that country was Britain. There was clearly a need for ships to communicate with their base and with each other. If wireless could do that, there would be a ready market.

The Young Marconi

The story of the invention of wireless has been told in countless books: for experts, for the layman, and even for children. It is a story of success. A young Italian hardly out of school brings to England the fruits of his research. He takes the country by storm, becomes rich and famous, and receives the Nobel Prize, the highest scientific accolade, at the age of 33.

Getting the Message: A History of Communications. Second Edition. Laszlo Solymar, Oxford University Press (2021).
© Laszlo Solymar. DOI: 10.1093/oso/9780198863007.003.0006

The name of the young Italian was Guglielmo Marconi. He had a young and beautiful Irish socialite for a mother and a rather morose Italian country squire for a father. The young man had no formal education; he was never much good at academic subjects. The motivation to work on radio waves (called Hertzian waves at the time) had apparently come from his reading an obituary when Hertz died in 1894. Marconi was then 20 years old. He became so interested in the subject that he started to do experiments himself. His laboratory was in the loft of his family's country house, and he used the estate as his experimental ground. It was not particularly difficult to start the experiments. He could build the apparatus himself from published accounts, and he was probably able to borrow some equipment from his neighbour, Professor Righi, who did actually lecture on Hertzian waves at the University of Bologna.

What was the state of the art at the time when Marconi started his experiments? It had been known then for seven years that waves that propagate in air can be generated and can be detected. The distance between the transmitter and the receiver was typically a few metres, which is convenient for demonstrating in a laboratory. Marconi's idea was to set the transmitters and receivers far apart, and use them to signal from one place to another. How original was Marconi's idea? Not very. Anybody who had known about the existence of Hertzian waves could have come to the conclusion that those waves might one day become useful for communications. To those in the know, one could add all the readers of the *Fortnightly Review*, or at least those who read Sir William Crookes' article in the February 1892 issue entitled 'Some Possibilities of Electricity'. He made the point very clearly:

Rays of light will not pierce through a wall, nor as we know only too well, through a London fog. But the electrical vibrations of a yard or more in wavelength of which I have spoken will easily pierce such mediums which to them will be transparent. Here then, is revealed the bewildering possibility of telegraphy without wires, posts, cables, or any of our present costly appliances... Any two friends living within the radius of sensibility of their receiving instruments, having first decided on their special wavelength and attuned their respective instruments to mutual receptivity, could thus communicate as long and as often as they pleased by timing the impulses to produce long and short intervals on the ordinary Morse code.

So there must have been hundreds, probably thousands, of people who could have been fired with the ambition of wanting to realize wireless telegraphy. As it happened the fire of ambition touched Marconi only. With dogged determination he set out to improve his apparatus. Could that qualify as a pursuit of science? Hardly. He was not a trained scientist and he was not guided by any desire to prove theoretical predictions. What he did was engineering pure and simple, using the time-honoured method of trial and error.

In 1896, when he was able to communicate for a distance of well over a mile, he judged the time ripe for bringing his invention to England. He arrived in the company of his mother who had good connections in London society. He wrote first to the War Office but when the reply was long in coming he succeeded, thanks to his mother's contacts, in obtaining an introduction to William Preece, the Chief Engineer of the Post Office. We met him in Chapter 5 as the adversary of Oliver Lodge, the villain unwilling to believe in the beneficial effects of self-induction. Concerning wireless, was he a hero or a villain? He could have been a great hero by supporting the unconventional ideas on light and electromagnetic waves just taking shape in England and coming from the disciples of James Clerk Maxwell. It would have perhaps been too much to expect him actively to help his professional enemies. That would have required selflessness and magnanimity few chief engineers are endowed with. It would, however, not be improper to regard him as a minor hero who was at least able to recognize a promising device when he saw one, and was imaginative enough to foresee a potential market.

In spite of his youth, Marconi looks quite an imposing figure in the photograph (Fig. 6.1) taken soon after he arrived at England. He was clearly a man to be reckoned with. Preece had the same opinion. After their first meeting he immediately offered the help of the Post Office in arranging demonstrations. The first one of this series of demonstrations was held in July 1896 between two Post Office buildings in London. Its success was followed by a bigger demonstration in September on Salisbury Plain, where a distance of 1¾ miles was reached. With some further

Fig. 6.1 Guglielmo Marconi shortly after his arrival in England in 1896.

improvements in his apparatus Marconi succeeded in sending signals a distance of 4½ miles in March 1897 and 9 miles two months later.

Publicity

Marconi was more than a resourceful inventor. He turned out to be a master of publicity as well. After his first successful experiment he was never far from the public eye. Perhaps his greatest coup was reporting the Kingston yacht races in the summer of 1898 for the Dublin *Daily Express*. He set up his transmitter on the *Flying Huntress*, a tug hired for the races. The *Express*'s sailing correspondent made his expert comments from the bridge of the tug which were then speedily sent by Marconi's wireless telegraphy apparatus to the shore at Kingston, whence they were telephoned to the newspaper's Dublin offices. Marconi, as so often in his life, was lucky. The traditional way of reporting the races, watching the manoeuvres of the yachts from the shore by binoculars, ran into difficulties when heavy fog descended on the second day. Only a reporter close to the event, and the one whose message was transmitted by wireless, could continue his account.

The Dublin *Daily Express* devoted nearly two pages to the races endeavouring at the same time to describe Marconi's apparatus in great detail, and claiming some share of the glory for the Irish nation in the person of Marconi's mother. The leader written at the conclusion of the races was not only full of praise for the young inventor but it also paid tribute to the wonders of learning in general:

Science which robbed us of the fairy tales of imagination had dowered us instead with its own fairy tales of fact. There is indeed an essential poetry in the higher speculations and accomplishments of science[1]

Marconi's next exercise in public relations came a little later in the same summer. The immediate reason for the request of a wireless set was a knee injury suffered by the Prince of Wales. He wanted to be close to his mother, who resided at the time at Osborne House in the Isle of Wight, but not too close. He chose the royal yacht moored a couple of miles away in Cowes Bay. The wireless communications link set up by Marconi between Osborne House and the yacht was to the satisfaction of both the Queen and the Prince of Wales. The daily messages sent bore witness to the usefulness of the service provided by Marconi. The first one from the yacht was affectionate:

The Prince of Wales sends his love to the Queen and hopes she is none the worse for being on board yesterday.

The Queen was rather anxious. She enquired:

The Queen wishes to know how the Prince slept; how he is this morning...

[1] It is somewhat ironical to note that more than a century later, when science and engineering combined have delivered a lot more than wireless telegraphy, such praise is no longer the order of the day. The achievements of science are not likened any more to any kind of poetry. Why? Maybe poets are jealous that computers will soon write better poetry.

The doctors of the Prince of Wales reassured her:

His Royal Highness the Prince of Wales has passed another excellent night and is in very good spirits and health. The knee is most satisfactory.

There was only one incident marring the harmony between the royal household and the inventor of wireless telegraphy. Marconi wished to inspect the aerial set up in the grounds of Osborne House. While attempting to do so he was warned by a gardener to keep away because the Queen's privacy was not to be violated. Marconi's refusal to be bound by court etiquette was reported to Her Majesty who promptly replied: 'Get another electrician'. However, she relented later when it was explained to her that sacking the electrician was not practicable in the circumstances.

The main aim was again publicity when Marconi decided to set up a wireless telegraphy link between England and France in March 1899. Technically, it was not a difficult task to accomplish. By that time ships could communicate with each other over larger distances. Nor was it a commercial proposition because the existing submarine lines across the Channel could amply satisfy all the demand. But as a publicity exercise it was an unqualified success.

By the autumn of 1899 Marconi was back reporting yacht races, although this time in America, at the request of the *New York Herald* and the *Chicago Times.* He set up his apparatus on a steamer from where the yachts approaching the finishing line could be clearly seen. The news flash that the American *Columbia* was going to win against Sir Thomas Lipton's *Shamrock* made Marconi a celebrity on the other side of the Atlantic as well.

Marconi's Progress in Business

Marconi was more than a resourceful inventor and a master of publicity. He was a shrewd businessman as well. After patenting his invention in July 1896, he turned his attention to business matters. With the aid of his cousin, who had good contacts in both scientific and financial circles, he managed to found the Wireless Telegraph and Signal Co. Ltd (it changed its name to Marconi's Wireless Telegraph Co. Ltd two years later) in July 1897, with capital of £100,000. He received £75,000 for his patents and for his services, leaving working capital of £25,000.

Now Marconi was famous and had the capital to produce wireless sets; all he needed was to find customers. His first customer, as may be expected, was the Italian Navy, anxious to reward the prodigal son. The British Navy was similarly inclined but decided to do so only after rigorous tests under simulated battle conditions. During the 1899 naval manoeuvres the flagship *HMS Alexandra* and the cruiser *Juno* were provided with Marconi's sets. Communications could still be maintained between them when fog descended. A journalist of *The Times* who followed the events from close quarters reported:

Our movements have been directed with an ease and certainty and carried out with a confidence which, without this wonderful extension of the range of signalling, would have been wholly unattainable. It is a veritable triumph for Signor Marconi.

Their lordships of the Admiralty were equally impressed. They wanted the sets at the right price. Marconi demanded the Earth and he very nearly got it. After prolonged negotiations the Admiralty ordered in July 1900 six coastal and twenty-six ships' sets for £3,200 each, to be supplemented by an annual royalty of £100 a set. The relationship between Marconi and the Admiralty was, however, not always harmonious. When it came to commercial acumen the Admiralty was not far behind. They sent one of their recently acquired sets to Ediswan, another company working in the field, and had 50 copies made.

The orders from the Italian and British Navy certainly helped the finances of Marconi's company but that would not have been enough to make the company profitable. The navies of other major countries, Germany, France, and the United States, showed little interest. They were nationalistic enough to want to develop their own industries. The logical choice for Marconi was to find customers in the merchant navies. His stroke of genius (as Bell's and Hubbard's before him in the telephone business) was to lease the apparatus and not to sell it. For that purpose in April 1900 he set up the Marconi International Marine Communications Company, which was to operate a chain of shore stations and provide each subscribing ship with a set and a wireless operator. The major breakthrough of the company was a contract with Lloyd's signed in September 1901. Lloyd's had a worldwide network of marine insurance whose functioning was greatly assisted by the possibility of keeping in contact with all ships on the high seas.

The next stroke of genius was the non-intercommunications policy. All Marconi operators had strict instructions not to communicate with the wireless operators of any other company except in case of emergency. Together with the company's head start in business, this policy clearly led to monopoly. A shipping company wanting to equip their ships with wireless sets hardly had a choice. If they wanted to communicate with shore stations they had to choose Marconi's apparatus. For any competition to have been successful they would have had to duplicate the world-wide network already in existence. Thus, Marconi managed to acquire customers from other countries too, such as North German Lloyd, Compagnie Transatlantique, the Canadian Beaver line, and the Belgian mailships. To remedy the situation the German government invited the interested nations to the First International Radiotelegraph Conference in 1903. All the participants, with the exception of Italy and

Britain, voted in favour of intercommunication. Britain ignored the resolution. Three years later at the second conference, held again in Germany, 21 countries were in favour of marine intercommunication, and six, Britain among them, were against it. The issue was finally resolved only at the third conference, held in London in 1912, in the wake of the *Titanic* disaster. It then became obvious that for marine safety, intercommunication was an absolute necessity.

Frequency and Wavelength

Looking at it from a communications viewpoint the most important property of an electromagnetic wave (the family to which radio waves belong) is its frequency. The concept of frequency has already been introduced in Chapter 5 when discussing the properties of sound. It is the same thing as pitch, and everyone knows what pitch is. In today's world of hi-fi enthusiasts many people would also know the range of frequencies audible to the human ear. It extends from about 20 to 15,000 Hz (15 kHz).[2]

Another characteristic of a wave is the wavelength.[3] It is mentioned less often than frequency. The reason why it must be included here is mainly historical. The division of radio waves into long wave (LW), medium wave (MW), and short wave (SW) was originally based on wavelength. Long wave means long wavelength, and the long wavelength region is defined nowadays as extending from about 2 to 1.2 km. Medium waves, as may be expected, are shorter. They extend from about 187 to 575 m. And finally the range of short waves is from 17 to 43 m.

Wavelength and frequency are not independent of each other. They are just two different ways of describing a wave. If one is known, the other one may be determined from the known velocity of the wave. For radio waves this velocity is 3×10^8 m/s = 300,000,000 m/s, which is actually true for all electromagnetic waves.[4] Then the frequency (denoted by f) can be determined from the wavelength by the simple relationship

$$f = 300,000,000 \text{ divided by the wavelength measured in metres.}$$

Hence the frequency bands in the LW, MW, and SW regions may be obtained as 150 to 250 kHz for LW, 520 to 1600 kHz for MW, and 7 to 18 MHz for SW (see Fig. 6.2). These are the scales that can be found on modern radio sets. In fact, short wave scales are usually missing because they are mainly used nowadays for receiving news from faraway countries for which the current demand is rather small. On the other hand, there is a new band from 88 to 104 MHz, which is sometimes referred to as the ultra-short wave region but which is usually designated as VHF, standing for very high frequency.[5]

[2] The symbol k stands for one thousand. It is a familiar sight in units like km and kg. Similarly, one may introduce M for a million (pronounced as Mega) and G for a thousand million (pronounced as Giga with both g's being hard).

[3] The easiest way to visualize wavelength is to imagine a water wave in a quiet lake caused by a pebble being thrown in. As the waves move away from the point of impact, one may see troughs and peaks. The distance between two neighbouring troughs or two neighboring peaks is the wavelength.

[4] 3×10^8 is pronounced as 'three times ten to the eight' meaning that there are 8 zeros after the number 3.

[5] Very high frequency is, of course, very far from being very high frequency. It is only very high frequency as far as radio waves are concerned. The spectrum of electromagnetic waves useful for telecommunications goes up to much higher frequencies.

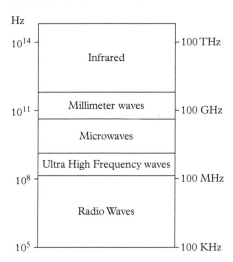

Fig. 6.2 The electromagnetic spectrum up to visible light.

The Scientific Background

The idea that a portable radio set, unconnected to anything electrical, can actually pick up some programme from thin air is still regarded with some wonderment. We can imagine how much more of a mystery it was a century ago. How could anyone ever think that such a thing was possible? Writers with vivid imaginations, like Jules Verne or H. G. Wells, took their readers to the world of rockets, flying ships, and space travel well before their time. However, obtaining a message without any visible connections was beyond the imagination of the most imaginative.

It is easy to say how it did not happen. It did not come about in response to consumer demand. Nor did the chief executive of a big telegraph company set up a research group and urge them to invent electromagnetic waves. It was not a serendipitous discovery either. No one stumbled across electromagnetic waves lying unnoticed on the street. The idea was born exclusively in James Clerk Maxwell's mind after a long process of germination.

It all started with Michael Faraday's lines of force, which he imagined as extending from one electrically active body to another, filling the space in between. It was a revolutionary concept. How did Faraday think of it? Why was he the only one who thought in those terms? Because he was a unique type of physicist, quite different from the nineteenth-century prototype. The prototype was an excellent mathematician who always endeavoured to describe new physical discoveries in mathematical form. Faraday never had any formal education. He was in the book-binding trade from the age of 12 to 21 when, thanks to his private studies in physics, he succeeded in becoming a rather humble assistant (sometimes undertaking the role of a valet as well) to Humphrey Davy of the

Royal Institution in London. Since Faraday knew no mathematics he had to live by the wits of his physical intuition. His experimental results were admired but his theoretical views did not command much respect among his contemporaries. It was believed that the theories of 'action at a distance',[6] which had been prevalent since Newton, could well explain all the phenomena found experimentally.

Faraday's lines of force might have been forgotten, along with many other theoretical artifices postulated at one time or another, had not the problems of signal distortion arisen in the practical field of telegraphy. The qualitative explanations provided by Werner von Siemens and Faraday, and the mathematical analysis of Thomson, all pointed in the direction of the space between the conductors having a significant influence upon the behaviour of the system.

Why did Maxwell become interested in Faraday's ideas? It was probably the intuition of a genius. He must have seen there a chance for further development. Maxwell offered his piece of work under the title 'On Faraday's Line of Force' to the Cambridge Philosophical Society meeting in December 1855, at the age of 24. What he did was nothing particularly original. He did what Faraday could not do. He put the idea of the 'lines of force' into mathematical form, the result being a set of equations connecting the electric and magnetic quantities to each other. Having done this he proved that all the results which had been derived using 'action at a distance' theories could also be obtained from Faraday's formulation. The theory, focusing attention on the space between the conductors, suddenly became respectable.

Maxwell returned to the problems of electricity six years later. He looked at his set of equations which could account for all electrical phenomena known at the time. He noticed then that if he introduced one more term to the equations, which he called the displacement current, then some new results emerged. The mathematical solution showed the existence of waves which propagated with a velocity very close to the known velocity of light. So he took the next logical step and claimed that light must be an electromagnetic wave. The paper was read in December 1864.

Even today, the imagination still boggles when one is told that a set of equations which can describe the propagation of light can also account for the deflection of a magnet in the vicinity of a current-carrying wire. How can it be possible? Aren't those things entirely unrelated? The sign of a great theory is just that: that it makes previously unrelated quantities suddenly fit into a common pattern. Still, at the time the whole thing looked very improbable. As improbable as if a biologist had come to the conclusion that the eating habits of giraffes could be described using the exact same principles as those used to describe the mating habits of mice.

[6] Meaning, for example, that the Earth, a certain distance away, could act upon the Moon via a gravitational force.

Was it easy to be wise after the event? Having seen Maxwell's equations, did all physicists exclaim with joy, 'Yes, now we know all secrets of electricity'? They did nothing of the sort. They ignored all those equations. Not out of malice nor on account of incomprehension. It just seemed too fantastic. The majority believed that some kind of minor modification of the 'action at a distance' theory would provide all the answers. William Thomson belonged to that majority. A further difficulty with Maxwell's theory was that nobody could see a way of proving or disproving it experimentally. What's the good of a theory without experimental confirmation?

One may well appreciate the problems of experimental physicists at the time even if they were inclined to believe, with Maxwell, that light was an electromagnetic wave. That hypothesis did not open up any new experimental approaches. The wavelength of light was known to be in the range 0.4 to 0.7 μm (μ stands for one millionth; therefore, a μm is equal to one millionth of a metre or, equivalently, one thousandth of a millimetre). Using the relationship between wavelength and frequency, the lowest wavelength quoted earlier corresponds to a frequency of 750,000,000,000,000 Hz. There was at that time no chance whatsoever of creating electrical quantities varying at that rate.[7] So the best thing was to wait and see.

For a long time nothing happened. Then some rival theories using different hypotheses appeared. In the year 1880 Hermann von Helmholtz, a professor at the University of Berlin, thought the time was ripe to attack the problem. He had a bright young post-doctoral student by the name of Heinrich Hertz. He asked him to prove or disprove Maxwell's theory.

It was already known at that time how to produce electrical currents varying fairly fast.[8] George Francis Fitzgerald showed that according to Maxwell's theory these currents should give rise to electromagnetic waves of the same frequency. So if the theory was correct there was a way of exciting electromagnetic waves: but how could one find out whether the excitation was successful? How could one detect the presence of electromagnetic waves? There was no known method at the time. Hertz was aware of the difficulties but accepted the challenge. Although he could not devote all his time to the problem—he had to do other things too to make a living—he kept on thinking about it and doing a variety of related experiments. By 1888 he had the answer. He could produce electromagnetic waves with a very simple apparatus which consisted of a coil, two big metallic spheres connected by a piece of wire, and a gap between two small metallic spheres (Fig. 6.3). The gap was 7.5 mm, the big spheres had diametres of 30 cm, and the wire between them was 3 m long. Hertz discharged a coil across the gap. The result was a fast varying current in the wire with sparks jumping across the gap.

[7] They can, of course, be created nowadays. That's what lasers do.
[8] About a million times slower than that needed to excite light, but that is still pretty fast.

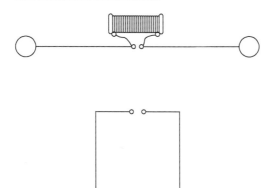

Fig. 6.3 Schematic representation of Hertz' transmitting and receiving apparatus.

The problem of detection was solved at the same time. Hertz noticed that when he produced a current in his apparatus, sparks also appeared across the gap of a similarly constructed device. Detection was via the gap in which sparks appeared when an electromagnetic wave was present. The simplest form of Hertz' detector is also shown in Fig. 6.3. It just consisted of a square of copper wire provided with a small gap. The key to the experiment was the appearance of sparks. Hertz published his findings in two papers which quickly followed each other. He announced proudly that he had had reception even when the two circuits were a great distance from each other. By 'great distance' he meant a few metres.

Hertz's discovery was greeted with great enthusiasm by the Maxwellians, Lodge, Heaviside, and Fitzgerald, who took up the torch when Maxwell died in 1879. Lodge repeated Hertz's experiments in England both with a spark detector and, later, with an improved detector which was known as a 'coherer'. There were lots of other people doing experiments on electromagnetic waves around the same time, e.g. Professor Righi in Italy, Edouard Branly in France, Ernest Rutherford in New Zealand, Capt. H. Jackson in England, Aleksandr Stepanovich Popov in Russia, and Nikola Tesla in the United States. Did anyone think of commercial applications? William Crookes, as quoted earlier, recognized the potential of electromagnetic waves for telegraphy but did nothing about it. Was there anyone else? Yes, Rutherford. He arrived in Cambridge in the autumn of 1895, bringing with him his invention, a detector of electromagnetic waves based on some magnetic effects. He left a fiancée behind at the other end of the world. In order to marry her he needed money. In order to get money he was thinking of cashing in on the commercial applications of wireless telegraphy. He wrote to his fiancée in early 1896:

I have every reason to hope that I may be able to signal miles, without connections, before I have finished. The reason I am so keen on the subject is because of the practical importance. If I could get an appreciable effect at ten miles, I would probably be able to make a considerable amount of money out of it, for it would be of great service to connect lighthouses and lightships to the shore, so that signals can be sent at any time. It is only in an embryonic state at present, but if my next weeks' experiments come out as well as I anticipate, I see a chance of making cash rapidly in the future.

It is not known how much effort Rutherford put into his electromagnetic experiments at Cambridge. Probably not much because the head of the Laboratory, J. J. Thomson (no relative of William Thomson), persuaded him to work on his own pet project, gas discharges. J. J. Thomson went on to discover the electron, and Rutherford became the founder of atomic physics.[9]

It seems quite likely that many scientists thought of the possibility of wireless telegraphy but for one reason or another they never made a serious attempt. They might have been more interested in the greater glory of science or perhaps they were deterred by the prospect of a long slog ahead. When an enthusiast like Marconi came along he had a clear run.

Let me finish this section by saying a few more words about James Clerk Maxwell. It is no exaggeration to say that his genius changed the course of physics. He made a brilliant mathematical hypothesis and he turned out to be right. When the theory of relativity came along, Newton's equation needed modification but not Maxwell's.

The significance of Maxwell's contributions was, of course, widely recognized as soon as Hertz provided experimental proof. One of the great men of physics, Ludwig Boltzmann, spared no praise when he lectured on Maxwell's theory in Munchen in 1893. He invoked Goethe, giving as his motto a slightly paraphrased version from Faust's monologue at the beginning of the play,

War es ein Gott, der diese Zeichen schrieb,
Die mit geheimnisvoll verborg'nem Trieb,
Die Kräfte der Natur um mich enthüllen,
Und mir das Herz mit stiller Freude füllen.

which may be rendered into English as

Was it a god that fashioned this design,
Which with a secret thrust divine,
Makes Nature's powers about me manifest,
And fills my heart with quiet happiness.

Some lecturers in our own time (myself included) are also in the habit of introducing the subject of electromagnetic theory by reference to divine action:

In the beginning God created the heaven and the earth and the earth was without form and void and darkness was upon the face of the deep and the spirit of God moved upon the face of the waters. And God said

[9] It would be interesting to speculate how the science and art of radio would have developed had Rutherford devoted his life to it. One might argue that the whole electromagnetic spectrum, from long waves to visible light, would have been more rapidly explored, leading to the early appearance of radar, microwave communications, and, possibly, the compact disc. The clear loser would have, of course, been nuclear physics. On the whole, would it have been a good thing or a bad thing for mankind? Even with hindsight it is difficult to tell what people would have achieved had they chosen one particular path in preference to another one. Take Shakespeare, for example. He could have become the greatest poet the world has ever known had he not got mixed up in show business.

$$\nabla \times H = J + \frac{\partial D}{\partial t} \qquad \nabla . B = 0$$

$$\nabla \times E = -\frac{\partial B}{\partial t} \qquad \nabla . D = \rho$$

and there was light.

Finally, I wish to quote the views of Richard Feynman, one of the greatest physicists of the previous century, who said in his lectures that

From a long view of the history of mankind—seen from, say, ten thousand years from now—there can be little doubt that the most significant event of the 19th century will be judged as Maxwell's discovery of the laws of electrodynamics.

The Principles of Radio

Hertz's apparatus was shown in Fig. 6.3. It worked. It was the first one that worked; nonetheless it is not the best for explaining the underlying principles. From a more modern viewpoint, an apparatus that will radiate electromagnetic waves may be represented schematically by the block diagram of Fig. 6.4.

Obviously, there must be a source of power: you can't get something out of nothing. The next element is an oscillator. Its function is to produce a fast-varying current as shown in Fig. 6.5a. If it just goes on indefinitely it can't possibly carry any information. The information is added by the modulator which, for wireless telegraphy, causes the current to be switched on and off, yielding, for example, the output shown in Fig. 6.5b. The fourth element in the diagram is the transmission line needed to lead the current to the aerial, and the aerial is there to radiate the power out. In Hertz's experiment there was no modulator (he had no intention of transmitting any information) and there was no clear distinction between oscillator, transmission line, and aerial, but later all these components became separate.

Hertz's spark-transmitter radiated out a wide range of frequencies. That was not suitable for communications. For communications modulated single-frequency output, like that shown in Fig. 6.5b, was required. This way conversations between ships, even if overlapped, did not affect each other, provided they were at different frequencies.

Fig. 6.4 A block diagram of a radio transmitter.

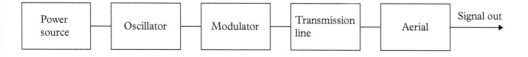

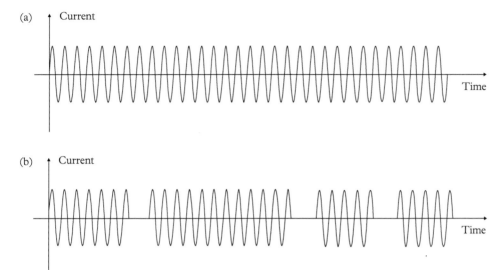

Fig. 6.5 (a) Continuous alternating current; (b) pulsed alternating current.

Signalling across the Atlantic

Marconi managed to stretch the distance over which he could communicate further and further. The way he achieved this was to have more powerful transmitters, more sensitive receivers, and bigger and better aerials. The array of masts built for experimental purposes at Poldhu in Cornwall is shown in Fig. 6.6. These were obviously elaborate structures which could not be built on ships. So the crucial experiments had to be done between ground stations. Marconi did indeed succeed in communicating from Poldhu to Crookhaven in Ireland, a distance of over 200 miles. The experiment convinced Marconi that the curvature of the Earth would not limit the achievable distance. He believed that radio waves, if launched by a sufficiently powerful transmitter, would propagate through the Earth, and was ready to undertake the great leap to communications across the Atlantic.

At first his usual luck seemed to have deserted him. A gale destroyed the aerials in Poldhu (Fig. 6.7) and at nearly the same time the aerial at Cape Cod, on the other side of the Atlantic, also collapsed as if in a conspiracy to doom the experiment. Marconi took it all in his stride. He had a less elaborate aerial erected quickly at Poldhu, and chose a site for the receiving station much closer to Poldhu in Canada. To support the aerials he relied on kites and balloons. Against all the odds he succeeded. On 11 December 1901 he clearly received three dots, the Morse signal for the letter S—or at least that's what he claimed. He could not repeat the experiment, partly because of the weather and partly because he was warned by the solicitors of the Anglo-American Telegraph Company that communications from England to Canada was the monopoly of their client.

Fig. 6.6 The aerial at Poldhu built for transatlantic transmission.

Fig. 6.7 Aerial destroyed by a storm before it could be used.

Marconi's claim was widely disbelieved. To prove his critics wrong Marconi returned to England and sailed immediately back again on the same ship, taking this time his receiving apparatus with him on the ship and persuading the ship's company to add a bit to their mast in order to accommodate a 150-ft aerial. While sailing to America he was able to get complete messages from Poldhu up to a distance of 1,550 miles, and the letter S could be received up to about 2,000 miles. Marconi's triumph was complete.

Why was so little credit given to Marconi's first claim to have spanned the Atlantic? The evidence was no doubt a little tenuous. To have received just three dots of the letter S was not a very strong claim. Secondly, the leap seemed too big. Just because it was possible to communicate over 200 miles it did not immediately follow that 2,000 miles would also prove to be possible. In the third place, none of the respectable scientists thought it possible to send electromagnetic waves across the Atlantic. It followed from the existing theories that electromagnetic waves would either shoot out into space (like light) or, at lower frequencies, follow the curvature of the Earth to a limited extent. In either case the chances of transatlantic communications were deemed to be zero.

So why were all the scientists proved to be wrong by Marconi's success? Did they misuse Maxwell's theory? Or was it that after all Maxwell's theory did not have the universal validity its supporters claimed? No, the scientists were right. Marconi should not have been successful. As it turned out later the reason for his success was the existence of an electrically charged layer in the atmosphere, the existence of which nobody suspected at the time.[10] The transmission was due to the fact that the signal propagated between two conductors: the sea and the electrically charged layer high up. Marconi was extremely lucky. Luckier than ever before. Had he failed in his attempt it would have seriously affected his credibility and also his business prospects, resulting possibly in bankruptcy. As it happened Marconi could move on to greener and greener pastures.

The Rise of Germany

In the first half of the nineteenth century Germany was regarded as the country of poets and philosophers. It was not an industrialized country. British and French engineers tended to look down at their German counterparts. The country was quiet, divided into lots of principalities. Perhaps a little surprisingly, revolutionary fervour caught on during the heady years of 1848–9. Less surprisingly a reactionary backlash followed when all the concessions were withdrawn. However, quite soon afterwards, a political consensus was reached incorporating a fair number of liberal ideas. The country was ready for a great leap forward. Two typical measures of German industrial progress (output of pig iron and

[10] Marconi's success in sending messages across the Atlantic made scientists think about the possible mechanism. The idea of an electrically charged layer in the atmosphere as a possible explanation came from Heaviside. He argued that the rays of the Sun might be able to 'ionize' a gas, i.e. create a negative electron and a positive ion from a neutral atom. In acknowledgment of his role in postulating this layer it is often called the Heaviside layer, although nowadays the term ionosphere is more widely used.

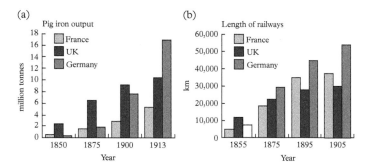

Fig. 6.8 The progress of Germany as shown by (a) pig iron output and (b) length of railways built.

length of railways), relative to that of France and Britain, may be seen in Fig. 6.8. Germany became a power to be reckoned with.

In the manufacture and laying of cable Germany had no interest. She had only a few colonies: there was no point in building a global network. However, wireless was in a different category. The motivation to communicate with ships was as strong for Germany as for Britain, but, in addition, wireless was important to Germany for communicating with the rest of the world in case of war. So Germany was interested from the very beginning. Professor Adolf Slaby of the Technische Hochschule in Berlin witnessed the experiments conducted by Marconi on Salisbury Plain. He returned home determined to set up a German radio industry. He developed a system in collaboration with Count von Arco which did not infringe Marconi's patent. Production started at the giant Allgemeine Electrizitäts Gesellschaft (AEG).

A rival development occurred at the University of Strasbourg under the direction of Professor Karl Ferdinand Braun, already a major figure in German science.[11] The production facilities were provided by Siemens & Halske. Under pressure from the German government these two interests merged under the name of Telefunken, which had a practical monopoly in Germany and was quite aggressively seeking markets both in Europe (they provided equipment for the Russian Navy) and in North America.

In 1906 Telefunken began building a station at Nauen. The aim was to communicate both with Togo, a German colony in Africa, and with the United States. In 1911–12 they erected aerials which covered as much as two square kilometres. With a transmitter capable of radiating 200 kW they had the most powerful machine in the world at the time.

The beginnings of radar could also be traced to the Germany of that period. The first determination of the position of a metallic object by reflection of radio waves was made by Christian Hulsmeyer. In 1904 he was able to detect ships at ranges up to 3000 m. The device he developed (he called it a 'telemobiloscope') was for preventing collisions between ships. For some odd reason neither the navy nor any of the commercial firms were interested in his invention.

[11] Braun was the inventor of the cathode-ray tube (today's television picture tube), the first to observe the rectifying effect in semiconductors, and a pioneer of directional aerials. He shared the Nobel Prize with Marconi in 1909.

The US Scene

Marconi's first encounter with representatives of the American Navy in 1899 resulted in no orders. They wanted wireless sets but could not make up their mind whether they should manufacture them themselves or buy them at home or abroad. Commander Barber warned against Marconi's non-intercommunication policy. He wrote to Admiral Bradford in 1901:

Such a monopoly will be worse than the English submarine cable monopolies which all of Europe is groaning under and I hope the Navy Department of the US will not be caught in its meshes.

Considering that by 1904 navies on both sides in the Russian–Japanese war were well equipped with wireless sets, it is somewhat surprising that the Americans were well behind. Obviously, the main reason was the lack of any military threat. They were in no hurry. An additional factor was the fierce independence of commanding officers. Before the advent of the wireless the captain of a ship was its absolute ruler. With a wireless set on board he could receive commands from a superior officer and that was the last thing any of the captains wanted. Hence there was a stubborn opposition by the old guard to any attempt at curbing their freedom. In fact, it was not until 1912 that the American Navy was supplied with sets and the new flexibility was incorporated in military strategy and tactics. When the United States entered the First World War in 1917 the navy was there on equal terms with the European powers.

Commercially, the power of the American Marconi Company was enormous. They sued several of their competitors in America for patent infringement and put them out of business. They did, for example, acquire in this way 70 shore stations of the United Wireless Company. Nonetheless, they were regarded as a foreign company which could not be trusted when the security of the United States was at stake.

Murder

As mentioned in Chapter 4 telegraph wires became known to Londoners as 'them cords that hung John Tawell'. It would have been very unfair if wireless had not had a similar opportunity to rise to fame. It had. The man who committed the murder was called Dr Hawley Crippen. The story, as it unfolded, gripped newspaper readers on both sides of the Atlantic.

It all started with the disappearance of Mrs Cora Crippen in February 1910. The husband was interviewed several times by Chief Inspector Walter Dew of Scotland Yard, the last time on 8 July. By 11 July Dr Crippen had disappeared together with his secretary, a Miss Ethel Le Neve.

On 12 July Scotland Yard found a dismembered body in Dr Crippen's house. Next day, a reward of £250 was offered for information leading to the arrest of Dr Crippen.

On 20 July in Antwerp a Mr Robinson and his son boarded the good ship *Montrose*, bound for Quebec. The captain of the ship, an avid reader of newspapers, was well aware of the hunt for Crippen. He became suspicious when he saw the alleged father and son holding hands romantically. In the best traditions of Sherlock Holmes he also discovered telltale signs of spectacles worn in the past and of a recently removed moustache, both associated with Dr Crippen. The good captain reported his suspicions by wireless to the Canadian Pacific Railway Office in Liverpool who, in their turn, contacted Scotland Yard in London. Not wasting any time, Chief Inspector Dew took a fast ship to Canada. To the excitement of all newspaper readers (the progress in the Atlantic of both ships was daily reported), he arrived in Canada in time to board the *Montrose* in the St. Lawrence River. Mr Robinson and son were arrested on 31 July.

According to *The Times* (which devoted not less than four columns to the event)

There was something intensely thrilling almost weird, in the thought of these two passengers travelling across the Atlantic in the belief that their identity and their whereabouts were unknown, while both were being flashed with certainty to all quarters of the civilized world.

The article concluded with quiet satisfaction: 'escape no longer lies across the ocean'.

Disasters

Wireless telegraphy was pressed into service by the requirements of the military. Its main civilian function was to call for help when disaster struck. The first distress call by radio was made in 1899 by the East Goodwin Lightship to bring assistance to the stranded German ship *Elbe*.

The disaster that called attention to the capabilities of wireless communications more loudly than anything else was the sinking of the *Titanic* on its maiden voyage in 1912. It was at the time the largest liner in the world, provided with all the luxuries money could buy. On 14 April it struck an iceberg and sank within three hours. It was provided, of course, with a Marconi wireless set,[12] which was made good use of by the illustrious passengers who wanted to be kept up to date with news from Britain and America.

Reports of weather conditions were not as yet part of established routine; nevertheless, the *Californian* (which was in the vicinity) did send a telegram to the *Titanic* saying that she was stopped and surrounded

[12]The internationally accepted distress signal was CQD at the time, but SOS (on account of its distinctive feature in Morse code $\cdots - - - \cdots$) was already agreed to come into use from June 1912. The *Titanic* used both. The very last SOS signal in Morse code was sent out on the last day of 1997 by the Bahama-registered *MV Oak* en route from Canada to Liverpool. In any case it has not been much used in the previous decades because advances in technology made possible the much more suitable verbal contact.

Fig. 6.9 Tributes paid to Marconi.

by ice. The receipt of the message was acknowledged by the wireless operator on the *Titanic*, but he asked to be excused from further conversation on account of the numerous calls coming from America. Since it was late at night and the operator of the *Californian* had nothing better to do, he went to bed. Fortunately, there was another ship nearby, the *Carpathia*, whose operator happened to be less sleepy. He caught the distress signals broadcast by the *Titanic* and his ship moved as fast as it could into the disaster area, arriving two hours after the Titanic went down (the *Californian* was much nearer: it could have arrived before the *Titanic* sank). Of the 2200 passengers and crew, they managed to pick up about 700. The world outside did not have a clear picture of what happened for quite a while because communications were jammed by the great number of enquirers, including radio amateurs.

The crucial role played by the wireless was clear to everybody. Had the *Californian* had wireless operators on a 24-hour watch it seems probable that practically everybody could have been saved. The lessons were learned and new regulations (compulsory wireless sets on smaller ships, 24-hour watch on bigger ships, licensing of amateurs in a specified frequency band) were introduced.

The Times (16 April 1912) was full of praise for the new technology:

Imagination is filled once more by the wonderful part played by wireless telegraphy in the story of the *Titanic*...But for this new instrument of communication it might have been that the greatest product of naval architecture might have passed from our human ken, her fate for ever unknown. We recognize with a sense near to awe that we have been almost witness of a great ship in her death agonies.

There was clear evidence of better communications between ships a year later when the next disaster struck. The *Volturno*, carrying a full load of emigrants to the US, caught fire in the middle of the Atlantic Ocean and had to be evacuated. The distress call brought ten ships to the rescue. There was great public interest in the rescue operations since the conditions were rather dramatic: a blazing fire coinciding with a raging storm. The papers were full of praise for Marconi, whose sets were used. *Punch* published a congratulatory cartoon (Fig. 6.9a). Similarly inclined, the Guerin–Boutron Company included Marconi in its series of the benefactors of mankind (Fig. 6.9b).

Scandal

In the history of government contracts one may find instances when those near to the sources of power made some money thanks to their privileged positions. The Liberal government which came to power in

England in 1906 was all in favour of honesty. When the new Parliament met, Lloyd George proposed that 'members ought to have no interest, direct or indirect, in any firm or company competing for contracts with the Crown'. However, it turned out later that some of the Liberal ministers were not above suspicion themselves.

The company which was about to receive a government contract was Marconi's. The proposal, made in March 1910, was to build an imperial wireless chain of 18 stations at a cost of £60,000 each plus a royalty on the gross receipts. The tender was accepted by the Post Office in March 1912 and signed in July, but it still had to be ratified by Parliament. Meanwhile, in March, Godfrey Isaacs, the managing director of Marconi's, got involved with the stock exchange flotation of the American Marconi Company. He kept 100,000 shares for private disposal. Of those he sold 1,000 each to three leading members of the Liberal Party: the Attorney General who happened to be his brother Rufus; the Chancellor of the Exchequer, Lloyd George; and the Chief Whip, Lord Murray. When the terms of the contract became known, Marconi shares rose sharply on the stock exchange, and the shares of their American company also rose in sympathy. These three gentlemen made a handsome profit.

Soon afterwards, as a result of some investigative journalism, the *Eye Witness* accused the Isaacs brothers and the Postmaster General of enriching themselves at the expense of the British taxpayer. There was an uproar. Accusations of sleaze spluttered by indignant Tories in Parliament prevented ratification of the contract before the summer recess. Two Select Committees were appointed: one looking into the sleaze accusations, and the other one at the technical aspects of the contract. Both committees upheld the government's conduct. Marconi's was deemed to be the best supplier, and nobody was found to have used privileged information for making personal gain, although those involved from the Liberal Party were censured for their lack of frankness. In the debate on 18 June 1913 Isaacs had to declare: 'I say solemnly and sincerely that it was a mistake to purchase those shares'. Lloyd George made a similar statement: 'I acted thoughtlessly, carelessly, mistakenly, but I acted openly, innocently, honestly'. The scandal did not seem to harm either of them. Within less than six months Rufus Isaacs rose to become Lord Chief Justice of England under the name of Lord Reading, and Lloyd George went on to become prime minister in 1916. The only sufferer was Marconi's company. Although the contract was ratified on 8 August 1913 no substantial work was done before the outbreak of the First World War, and during the war period priorities were, of course, different. The contract had to be cancelled. It was, however, not all gloom for Marconi's. The company's suffering was somewhat alleviated by the sum of £600,000 it was awarded in damages against the Post Office.

The Second Industrial Revolution Gathers Pace: The Birth of Electronics

The interaction between physics, mathematics, and technology started with the work of William Thomson in the middle of the nineteenth century. It was his analysis of signal propagation on long submarine cables which broke the mould. From then on it was impossible entirely to ignore predictions based on abstract mathematical models. The 'men of experience' had to argue their views, and although they usually won the debates (see the discussion on Preece versus the Maxwellians in Chapter 5) they had to tread more and more carefully. The success of the periodic loading of telephone lines drove home the lesson that at least some of the advances would depend on sophisticated mathematical analysis. It should not, however, be assumed that mathematics played a role in all major advances. The greatest of all advances, the birth of electronics, came about in an entirely empirical manner.

The first experiments which could be qualified as electronics were carried out by Edison in 1883 as part of his attempts to find the right kind of filament for incandescent bulbs. He inserted in the bulb an additional metal plate and found that a small current flowed between the filament and the plate when the plate was attached to the positive end of a battery, but no current when the plate was negative. This became known as the Edison effect. Many scientists, among them Ambrose Fleming, repeated these experiments with similar results. Nobody understood what was going on until the discovery of the electron by J. J. Thomson in 1897.[13] After that it was slowly recognized that the heated filament (called the cathode) emitted negative electrons which could then proceed to the metal plate (called the anode) provided it was positive. If the metal plate was negative then the negative voltage repelled the electrons so that no current could flow.

Fleming realized that this simple device, which became known as a diode, could be used as a detector of electromagnetic waves if connected to a circuit which measures the average current. If the current flowing in the receiver in response to the incident electromagnetic waves is oscillatory, as shown for example in Fig. 6.5a, then the average current is zero. However, if the diode lets through the current only when the anode is positive, then the instantaneous current will take the form shown in Fig. 6.10 and its average will clearly be non-zero.

Fleming's diode made a significant difference to the art of detecting electromagnetic radiation but had electronics stopped there its contribution to communications would have been quite modest. The really great advance was made by Lee de Forest in the United States. He put into the diode a third electrode between the cathode and the anode. He called the device an *audion*, later to be called a *triode*. It rose to great significance as the main building block of an *amplifier*.[14] In fact, it

[13] The discovery is usually attributed to J. J. Thomson (no relative of William Thomson) mainly on the grounds that he was very active following up the discovery. There were, in fact, several groups on the Continent coming to the same conclusion (that a particle, much lighter than the hydrogen atom, exists) and, in particular, Emil Wiechert of Königsberg published his results before Thomson.

[14] A triode having three electrodes turned out not to be the ideal device for amplification. To improve its performance additional electrodes were needed, resulting in the tetrode, the pentode, the hexode, the heptode, and the octode. This led to the rather anomalous situation that, besides a select band of classicists, it was the electronic engineers who could count in Greek.

Fig. 6.10 Pulsed
alternating current after
rectification.

became even more important because an amplifier could be turned into
an oscillator by the judicious use of feedback. Thus suddenly a simple
and reliable solution was offered for both the generation and amplification of radio waves.

The birth of electronics completely changed the future potential of
communications. It provided three vital components, a detector, an
amplifier, and an oscillator. With those components it became eventually possible to communicate from any point on Earth to any other
point, whether with or without wires. But, of course, electronics did
more than that. Its use led to better measurement and control techniques in all industries which, in turn, accelerated the progress of science. To the claim that the Second Industrial Revolution started with
the triple interaction between physics, engineering, and mathematics,
the further claim can now be added that its continual progress was con-
tingent on the advent of electronics. Without electronics the Second
Industrial Revolution would have ground to a halt. With electronics in
its advance guard it became unstoppable.

Short Waves

Some time after the end of the First World War the question of an
imperial wireless chain arose again. The chain could have worked at
long waves, as it was envisaged before the war, requiring enormous aerials and needing an enormous amount of power. But Marconi surprised
the world once more. He proposed transmitters working at short waves
which needed much smaller aerials and only a few kW of power.

How did he do it? Oscillators were no problem thanks to the advent of
vacuum tubes. The main question was how far the signal would go. It
was known that short wave signals would *not* follow the curvature of the
Earth. They could only be used in line of sight like light. If the transmitter
and the receiver 'see' each other, then radio reception is possible; otherwise, not. That would surely mean that short waves are unsuited for
long-distance communications. However, it turns out that short waves
can span large distances for the same reason that Marconi was successful in sending signals across the Atlantic: the agent responsible for it is
the ionosphere, an electrically charged layer above the Earth. The difference was that in 1902 nobody knew about the layer, whereas two

decades later its existence (as hypothesized by Heaviside) was accepted although not yet proven.[15]

Marconi set out in his well-equipped private yacht, the *Elettra*, to measure the signals from the Poldhu short wave radio station. He sailed as far as 2,500 km away in the Atlantic and was, at least at night, always able to receive signals.[16] Measurements in other parts of the world showed conclusively that short wave communications could cover the whole Earth. The mechanism they relied on is shown in Fig. 6.11. The waves travelled by successive reflections between the conducting Earth and the ionosphere.

Now which is better, long waves or short waves? Starting off with the same amount of power at point A, which one will arrive with more power at point B? As may be seen in Fig. 6.12, long waves follow the curvature of the Earth, whereas short waves are reflected by the ionosphere. It turns out that a wave which sticks to the surface of the Earth is very lossy; hence, long waves will decline more. The second reason why short wave communications are more efficient is that the shorter the wavelength (i.e. the higher is the frequency), the more it is technically possible to concentrate the beam in a narrow direction. Hence, a larger proportion of the power arrives at the receiver.

By how much are short waves superior to long waves? By quite a lot. Transmitters need only 2% of the power, and of course if the power requirement is less, then everything is cheaper. A short wave station may cost as little as 5% of one built for long waves. So it made good sense for all new radio stations designed for communications over thousands of miles to operate at short waves.

[15]The proof was provided by Appleton as late as 1924.

[16]The difference in reception between day and night is due to the fact that the position and constitution of the ionosphere depend on the incident sunshine.

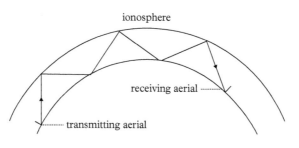

Fig. 6.11 The propagation of short electromagnetic waves between the Earth and the ionosphere.

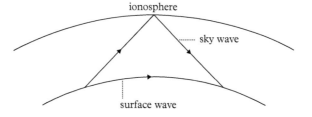

Fig. 6.12 Radio waves can propagate in two different ways, either by reflection from the ionosphere or by following the curvature of the Earth as a surface wave.

Negotiations with the government (the first ever Labour administration in the UK) were concluded by July 1924. The first link was to Canada, others following not long afterwards. The whole of the short wave Imperial Wireless Chain was in operation by the end of 1927.

Epilogue

The early development of radio was entirely dominated by one single man, Guglielmo Marconi. He was a first-class engineer, an expert in public relations, and a shrewd entrepreneur. There was nobody before him or after him who possessed such diverse talents. He loved power and money, the company of kings and queens, he collected lots of decorations all over Europe, and he had considerable influence, even in the Vatican. When he made the request, he had fewer difficulties in having his marriage annulled than Henry VIII.

His political choices, to say the least, were unfortunate. He joined the Italian Fascist Party in 1923, not very long after Mussolini's famous march on Rome. He became a member of the Italian Upper House, the President of the Royal Academy of Italy, and a member of the Fascist Grand Council, the supreme organ of the party. At the time of the Abyssinian War, he was Mussolini's roving ambassador, trying to lift the League of Nation's boycott against Italy. In contrast to his compatriot, Enrico Fermi, who left Italy in 1938 to participate in the American effort to build the atomic bomb, Marconi would have, without doubt, opted for the Axis Powers. As it happened, he did not have to make the choice. He died in 1937. At the news of his death all radio stations the world over observed a two minutes' silence, an honour not accorded to many inventors.

The Telephone Revisited

Introduction

The telephone made great strides ahead in a very short time. By 1911, as mentioned in Chapter 5, it was possible to make a call from New York to Denver, 4,300 km away, but obviously that was not far enough. Ambitions were much higher: firstly, to span the continent from New York to San Francisco, secondly to establish telephone connections between the US and Europe, and finally to enable people to conduct conversations between any two points on Earth.

What were the difficulties in achieving these aims? There were two main difficulties: attenuation (the telephone signals declining with distance) and the need to have a pair of wires for each conversation. The first one was a technological difficulty. In order to overcome that, amplifiers had to be invented. The second problem was perceived as an economic one. At what stage does it become unprofitable to erect additional lines?

Amplifiers made their appearance via the development of radio communications, the invention of the diode and the triode. For the telephone, to use a modern phrase, it was a technological spin-off. With the aid of amplifiers it turned out to be possible to reach San Francisco by 1915. AT&T managed to persuade the two ageing pioneers, Alexander Graham Bell and Thomas Watson, to open the line to New York with exactly the same words they used in their first telephone communication four decades before.

The requirement of a separate line for each conversation was a serious impediment to inter-urban communications. It took two further decades to solve that problem with the aid of carrier wave telephony and frequency division multiplexing. The fundamental principles will be described later in this chapter.

For Britain to establish telephone connections with the rest of Europe the technique of laying submarine telephone cables had to be mastered. For reasonable transmission, the telephone cable had to have a much greater diameter than that used for telegraphy. The first such cable was laid in 1891 across the Channel. It would have provided an excellent opportunity for French and British statesmen to talk to each other. The only such conversation was provided in the pages of *Punch* (Fig. 7.1). The French President's reference was to a fishing dispute in Newfoundland which had been poisoning Anglo-French relations for

Getting the Message: A History of Communications. Second Edition. Laszlo Solymar, Oxford University Press (2021).
© Laszlo Solymar. DOI: 10.1093/oso/9780198863007.003.0007

PRIVATE AND CONFIDENTIAL.

Lord Salisbury. "HALLO!"　　M. le Président. "HALLO!"　　Lord Salisbury. "YOU THERE?"　　M. le Président. "ALL THERE!"
Lord Salisbury. "CAN YOU SUGGEST AN *ENTRÉE* FOR DINNER?"
M. le Président. "*HOMARD AU GRATIN*,—AND, BY THE WAY, HOW ABOUT NEWFOUNDLAND AND LOBSTER QUESTION?"
Lord Salisbury. "NOT BY TELEPHONE, THANK YOU!!!"　　　　　　[*Telephone between London and Paris opened, Monday, March 23rd.*

Fig. 7.1

quite some time. The British prime minister, Lord Salisbury, did not think it was a subject that could be discussed on the telephone. The London–Berlin line was opened a good two decades later in 1913. Unfortunately, the idea of a hot line did not occur to anyone at that time either. Only *Punch* took the opportunity (Fig. 7.2) to make Lloyd George talk to the Kaiser on a subject close to the heart of both of them, taxation.

Submarine telephone cables to the Continent were laid without encountering any great problems, but to do the same thing across the Atlantic was a much more serious challenge. The main difficulty was to put amplifiers under the ocean. They needed long life. It would not have made sense to haul up the cable every month in order to replace a vacuum tube. The solution of the technical problems had to wait until 1956 when the first transatlantic telephone cable was laid, capable of carrying 35 telephone conversations. Did it mean that no transatlantic call was made until 1956? No, a regular service using radio waves had started nearly 30 years earlier.

THE GERMAN LLOYD.

Kaiser Wilhelm (*on the new Berlin-London telephone*). "HULLO, IS THAT THE CHANCELLOR? I SAY, WHAT DO YOU THINK OF MY NEW IDEA OF TAXING CAPITAL?"
Mr. Lloyd George. "EXCELLENT, SIR. MOST FLATTERING, I'M SURE."
Kaiser Wilhelm. "AND WHAT DO YOU DO WHEN THEY KICK?"
Mr. Lloyd George. "TAX 'EM ALL THE MORE."

Fig. 7.2

Wireless Telephony

The first man to transmit sound with the aid of radio waves was Reginald Fessenden of the United States, a radio enthusiast if ever there was one. He surprised the world, or at least all those who could listen to it, by broadcasting carols on Christmas Eve, 1906. Although the phonograph was invented by this time he preferred to give a live performance by singing himself. The quality of the singing might have been very high but reception was poor. The spark transmitters available at the time were unsuitable for sound transmission.

The situation changed considerably when oscillators became available. Then it proved possible for radio amateurs to talk to each other, or for one radio amateur to send out music for the entertainment of other radio amateurs. The idea of broadcasting gradually took root, culminating in the first commercial broadcast in Pittsburgh in

November 1920.[1] The best example of wireless telephony is, of course, radio broadcasting, but that is beyond the scope of the present book.

For communications on the high seas telegraphy was clearly superior to telephony. Messages between ships, or between ships and shore stations, were purely of a business character (apprehending Dr Crippen was an exception, although as far as Scotland Yard was concerned, that was business too) and it was better to have some written record for business transactions.

There was, of course, demand for intercontinental telephony. An early success was scored in 1915 with experimental speech transmission between the US Naval Station at Arlington, New Jersey and the Eiffel Tower in Paris. The first commercial telephone service between London and New York was inaugurated in January 1927 after collaboration between AT&T and the British Post Office. It worked in the long waveband. A short wave channel was opened in June 1928 with two further channels becoming available by the end of the following year. Other services followed quite quickly: in 1930 to Australia, in 1933 to South Africa, Canada, Egypt and India, and in 1935 to Japan.

The prices were rather high. A three-minute call from London to New York originally cost £15, reduced to £9 the year after.

The Principles of Sound Transmission by Carrier Waves: Modulation

Telegraphy worked by switching currents on and off. For telephony the current had to be modulated by the human voice, as was shown in Figs. 5.15 and 5.16. In carrier wave telephony the input sound modulates the high frequency current. The sound information is 'carried' by the wave: that's where the name 'carrier' comes from.

A pure, unmodulated carrier wave may be seen in Fig. 7.3a and a pure tone, a sinusoidal sound wave, in Fig. 7.3b. Two simple ways of affecting the carrier wave are amplitude modulation and frequency modulation, known widely by their acronyms AM and FM. As the name implies in AM the sound wave modulates the amplitude of the carrier wave (Fig. 7.3c). In contrast, FM leaves the amplitude unchanged but affects the frequency. The carrier wave frequency becomes higher when the sound is positive and lower when the sound is negative, as illustrated in Fig. 7.3d.

The role of the carrier wave is to carry the information along the line. When it reaches the receiver it needs to be demodulated; i.e. the original sound needs to be recovered. The process of demodulation for AM is shown in Fig. 7.4. First comes rectification which keeps only the positive part of the signal (Fig. 7.4a), and, secondly, the signal is fed into a device which retains only the slow variation, i.e. does away with the carrier wave. As a result the original sound wave is recovered (Fig. 7.4b). The

[1] Reporting the presidential election in which Wilson's Democrats were defeated.

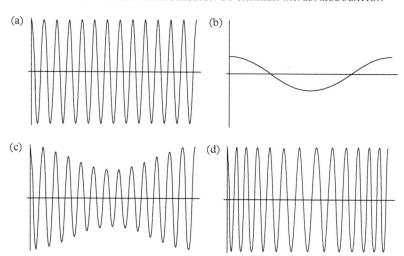

Fig. 7.3 (a) Continuous alternating current, (b) sound wave, (c) alternating current amplitude modulated by sound wave, and (d) alternating current frequency modulated by the sound wave.

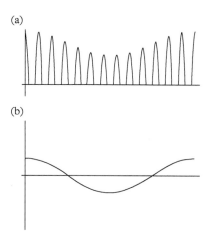

Fig. 7.4 (a) Amplitude modulated alternating current after rectification. (b) The demodulated sound wave.

device used for the demodulation of FM gives an output signal proportional to the frequency of the input wave, leading again to the recovery of the original sound wave.

The block diagram of a transmitter and a receiver suitable for telephony is the same as that shown for wireless telegraphy in Fig 6.4. At the transmitter an oscillator is needed to provide the carrier wave and a modulator to modulate the carrier wave by sound. At the receiver the wave coming in (i.e. the modulated carrier wave) must be demodulated.

There is one more very important concept not mentioned so far and that is bandwidth. The carrier wave is at one single frequency, say at 200 kHz. However, when this carrier wave is modulated there will be a range of frequencies present. For FM the relationship is quite complicated, but it is very simple for a particular kind of AM known as single sideband transmission. For sound frequencies going up to 3500 Hz (perfectly adequate for telephone conversations) the modulated carrier wave has frequencies from 200,000 to 203,500 Hz.

Frequency Division Multiplexing

Time division multiplexing was introduced in order to transmit several telegraph signals simultaneously on the same line. The trick was to intersperse the signals in time and then separate them at the receiving end (see Chapter 4). Frequency division multiplex does the same thing in the frequency domain. The information is put on a number of carrier waves of different frequencies which propagate entirely independently of each other, and then at the receiving end they are separated.[2]

How did the idea of frequency division multiplexing originate? It started in the telegraphy business which was big business in the 1870s. If by some means it was possible to send ten messages simultaneously on one single wire, that would lead to tremendous economies. Two men, Alexander Graham Bell and Elisha Gray, worked on the problem independently. Both hit on the idea of using tuning forks.

Tuning forks, as it was well known at the time, emit a well-defined sound frequency, or conversely they can be brought into vibration if sound of the same frequency is incident upon them. Thus, a possible way of sending a telegraph message is by interrupting the sound emitted according to the telegraphic code, and then converting the interrupted sound into electrical form. The remarkable thing is that, if one modulates a number of forks tuned to different frequencies in a similar manner, then they can travel on the same wire entirely independently of each other. When they arrive at the receiving end they can be separated according to their frequencies. It was a good idea but it just did not work out in practice. On the other hand once the idea of turning sound into electrical form took root, the obvious next step was to invent the telephone, and that's exactly what both Bell and Gray did.

The road to the practical application of frequency division multiplexing opened up when it became technically possible to (i) generate electromagnetic waves with well-specified frequencies, (ii) modulate them by voice, (iii) amplify them without distortion, and (iv) separate the different frequencies at the receiver. All these technical problems were solved by the middle of the 1930s. In 1938 the London–Birmingham trunk line was opened to service. It could carry as many as 40 simultaneous telephone conversations.

What will determine, at least in principle, the number of telephone conversations which can be carried on a single line? Is the answer the result of terribly complicated calculations which only the greatest experts can understand? The answer is no: for single-sideband AM transmission the calculation can be performed by anyone with some dexterity in the four basic mathematical operations. If one conversation

[2] Many people find it difficult to imagine that a large number of conversations may coexist on a single line without any effect upon each other. It may be worth pointing out that exactly the same thing happens with radio broadcasting. All the various programmes arrive simultaneously to our radio set without affecting each other (well, sometimes they do!) and the selection of the right station takes place in our own set by turning a knob.

is carried between 200,000 and 203,500 Hz then, leaving a little margin of 500 Hz, the next conversation could be modulated on to a carrier at 204,000 Hz, and the next carrier at 208,000 Hz, and so on. The number of telephone channels clearly depends on the available bandwidth. If a transmission line can carry frequencies from 0 to 25 MHz, then the number of telephone channels available is equal to

$$25,000,000 / 4,000 = 6,250.$$

Clearly, going up to 50 MHz would provide 12,500 simultaneous channels, and so on. Can this go on indefinitely? Can we have millions of channels? The answer is yes, but for that optical fibres are needed. They will be discussed in Chapter 12.

Publicity

In Europe, with the exception of Sweden and Denmark, the public did not easily take to the telephone. In some countries, particularly in France, it was regarded as an instrument of seduction. According to Colette, whose own moral standards were not particularly high, 'the telephone is only for men concerned with important business matters and for women who have something to hide'. The popular conception is well represented by the naughty postcard shown in Fig. 7.5.

For ladies of leisure, internal telephones were for easy communication with servants (Fig. 7.6). And of course the value of the telephone in

COME HOME ? NOT LIKELY, WHEN I'M HAVING SUCH A JOLLY FINE TIME

Fig 7.5

Fig. 7.6

Fig. 7.7 Promotion literature from the Post Office. Today, such encouragement to violent confrontation would be condemned as injurious to the human rights of burglars.

an emergency was always regarded as high, one of its main selling points (Fig. 7.7).

War

The advent of the telegraph radically affected the conduct of war by enabling the various commanders to keep in touch with each other. The telephone made only a minor difference. Nonetheless wars brought some new features to the use of the telephone which may be worth mentioning.

Air raids started in the First World War, permitting the civilian population to share the spirit of the front line. Two telephone operators at Whitley Bay, who kept on working during a Zeppelin raid in August 1916, received the medal of the Most Excellent Order of the British Empire for 'displaying great courage and devotion'.

The First World War also caused havoc in the French telephone system. It was caused not by the Germans, their adversaries, but by their allies the Americans, who entered the war in 1917. For them the telephone was a necessity. Finding the French network very much wanting, they set up their own lines. They installed 28,000 miles of wires and 273 exchanges and brought across the Atlantic a couple of hundred young ladies to look after them.

The installation of telephones in the Second World War was not particularly noteworthy except for one occasion. On 5 June 1944 Post Office engineers succeeded in laying a cable halfway across the Channel. The second half of the cable, right to the battlefield, was laid the day after, on D-day. One voice and a two-way teleprinter circuit worked by the afternoon. By D-Day plus one a carrier wave system using frequency division multiplexing was in operation.

The Beginnings of Telex

Telegraphy had always had the advantage over telephony that it provided a permanent record of the message. It had, however, two major disadvantages: the message arriving at the telegraph office still had to be delivered by a messenger, and secondly, skilled operators were needed both at the transmitting and at the receiving ends. Private lines connecting two particular offices had, of course, been in existence for a long time. The Universal Private Telegraph Company was founded in 1861 with exactly that purpose in mind. After private lines the next logical thing to do was to set up exchanges, but that had limited success. Only a few telegraph exchanges ever came into existence (e.g. Newcastle, Glasgow) and the number of their subscribers never reached three figures.

The situation changed for the better after the invention of the teleprinter by Charles Krum, an American, who like Almon Brown Strowger, was first involved in other professional pursuits. He was a cold-storage engineer. The marketing of an improved version of his invention started in 1918 with a firm called Morkrum, which was later taken over by AT&T. The main advantage of the teleprinter was that it dispensed with the need to synchronize the transmitter and the receiver. Letters were coded by five digits, as in Baudot's machine, but each letter was sent separately in the sense that it was preceded by a start signal and followed by a stop signal. There was no longer any need for skilled operators. The teleprinters worked automatically, switched on by the incoming signal. They very much resembled ordinary typewriters which any person in the office could operate. At the receiving end the teleprinter could print messages while entirely unattended.

By the 1930s there was a clear demand in Britain for a teleprinter network provided with exchanges. The name of the new service was made up by *tel* for teleprinter and *ex* for exchange becoming *telex*. The Post Office moved forward rather cautiously. Instead of setting up an entirely new network (as happened in Germany) they relied on the existing telephone lines for transmission and on the existing telephone exchanges for switching. In order to make the network believe that the signals were proper audio signals they modulated a voice frequency (first at 300 Hz and later at 1.5 kHz) with the telegraph signals. Inland service started in 1932. The first Continental link was established to the Netherlands in 1936 and extended to Germany in 1937.[3]

[3] Since Germany had a separate telex network, the messages sent from England had to be demodulated in Amsterdam and sent to Germany as telegraph signals.

On the eve of the Second World War Germany had the most extensive telex service in Europe. France had none. The reason may be that the French were too attached to the Baudot telegraph machine (a repetition of their earlier attachment to mechanical signalling which delayed their adoption of the electric telegraph).

Revolution, Civil War, and Occupation

In any well-planned revolution the first aim is to cut the communications of those in power and establish communication links for the insurgents. The best example is the October Revolution in Petrograd in 1917. John Reed, a left-wing American journalist, gave a sympathetic, and perhaps not entirely truthful, eyewitness account of how it was achieved.

The Telephone Exchange held out until afternoon when a Bolshevik armoured car appeared, and the sailors stormed the place...Tired, bloody, triumphant, the sailors and workers swarmed into the switchboard room, and finding so many pretty girls, fell back in an embarrassed way and fumbled with awkward feet. Not a girl was injured, not one insulted. Frightened, they huddled in the corners, and then, finding themselves safe, gave vent to their spite, 'Ugh! The dirty, ignorant people! The fools!'...Haughty and spiteful, the girls left the place. The employees of the building, the line-men and labourers—they stayed. But the switch-boards must be operated—the telephone was vital...Only half a dozen trained operators were available. Volunteers were called for; a hundred responded, sailors, soldiers, workers. The six girls scurried backwards and forwards, instructing, helping, scolding...So, crippled, halting, but *going*, the wires began to hum. The first thing was to connect Smolny with the barracks and the factories; the second to cut off the Duma and the junker schools.[4]

The Russian director Sergei Eisenstein went further. In his film version of the events the sailors started to dance with the operators. Trotsky, the architect of the revolution, gave a somewhat different account, less romantic and more to the point:

It had been assumed that the central Telephone Exchange would be especially well fortified, but at seven in the morning it was taken without a fight by a company from the Keksgozmsky regiment. The insurrectionists could now not only rest easy about their own communications, but control the telephone connections of the enemy. The apparatus of the Winter Palace and of central headquarters was promptly cut out.[5]

The civil war in Spain also gave rise to unorthodox uses of the telephone. In the chaos of the first days of Franco's military insurrection the telephone lines were still more or less intact. The adversaries were very often only a local call from each other. It gave an opportunity to use the telephone to exhort, persuade, threaten, or blackmail the opposite camp. A famous example, of which every visitor to Toledo's Alcazar had

[4] Smolny was the headquarters of the Bolsheviks, the Duma was a legislative body dominated by the moderate left, and the junkers were military cadets.
[5] The seat of the government.

been amply told until (and a bit after) Franco's death, is the call from the local militia to Colonel José Moscardó, the commander of the Alcazar. He was asked to surrender in exchange for the freedom of his son. He refused. The son was executed. The Alcazar, in a rather dilapidated state, was relieved by Franco's forces, two months later.

Another illustration of the significance of the telephone in the same civil war is provided by the events in Barcelona. There the attempted military takeover in July 1936 was foiled by armed anarchists, who represented a formidable force. Their heroic deeds in holding the front against their trained adversaries is well described in Orwell's *Homage to Catalonia*. By definition (after all anarchists prefer anarchy) they eschewed military organization, even refusing to wear a uniform. They had though a lot of influence in the trade union movement. In particular they controlled the telephone exchange. On 2 May 1937 a telephone call from Manuel Azaña, the president of the republic, to Lluís Companys, the president of Catalonia, was interrupted by the operator saying that the line was needed for more important purposes. The police response, to take over the telephone exchange, led to a civil war within the civil war. The anarchists and their allies were defeated by the combined forces of the Communists and of the government. The estimated casualties were 500 killed and 1,000 wounded.

Finally, we come to occupation. The rules and regulations introduced by the Nazis in the territories they occupied in the Second World War were never gentle and quite often, when their vital interests were at stake, unashamedly brutal. Communications was one of their main concerns. A 'warning' displayed at all post offices in occupied France may be seen in Fig. 7.8. It may be rendered in English as

Any damage caused to communications facilities (telegraph poles, cable apparatus, transmission equipment and post offices as well as to radio equipment) will be punished by death.

Head of Military Administration in France.

Regulation in the United States

What was the aim of the telephone monopolies in Europe? Not customer satisfaction—for state monopolies such considerations were entirely alien. Their main concern was to provide jobs for the boys in their own countries. Buying telecommunications equipment in another country was a rarity.

In the US the situation was more complicated owing to AT&T's near monopoly. Although their market share declined after the expiry of the Bell patents, they bounced back later by aggressive policies.

The wish to regulate AT&T started around the turn of the century, mainly in the southern states which wanted to encourage business

WARNUNG

Jede Beschädigung von Nachrichten-Anlagen (Drahtgestängen, Kabelanlagen, Vermittlungseinrichtungen und Postämtern, sowie Funkanlagen) wird mit dem Tode bestraft.

Der Chef der Militärverwaltung Frankreich.

AVERTISSEMENT

Tout endommagement de moyens de transmission (poteaux télégraphiques, jonctions de câbles, appareils, de bureaux de poste et d'installations radiotélégraphiques) est interdit sous peine de mort.

Le Chef de l'Administration Militaire en France.

Fig. 7.8 Warning displayed at post, telegraph, and telephone offices in occupied France.

expansion by ensuring low telecommunications rates. By 1914 thirty-four states and the District of Columbia had some kind of regulation. One of the problems that arose, one in which the anti-trust authorities also took an interest, was the denial of access to essential facilities by a monopolist. An example was a railway dispute in 1912. The complaint was that the bridges across the Mississippi were controlled by a few companies which denied access to their competitors. The US Supreme Court ruled that this was an unlawful restraint on trade. The implications for telecommunications were clear. Indeed, the US Department of Justice filed an anti-trust suit against AT&T which was settled out of court in 1913. AT&T undertook to interconnect with other companies for long-distance calls and to refrain from buying up competing telephone companies. After 1921 (when regulation passed on from anti-trust authorities to the Interstate Commerce Commission) AT&T returned to its predatory practices. By 1934 it owned 80 per cent of all the telephones and had full monopoly in all the long-distance services.

In order to remedy the situation the Communications Act of 1934 was passed, which set up the Federal Communications Commission. It was their job to protect the independents and to watch out for unfair practices. On the whole, however, they did little because they were believers in natural monopoly theory, which maintained that a monopolist could supply the market output at a lower cost than any combination of competing firms. This was largely a matter of belief because it would have been rather difficult to test the theory in the field. A monopoly was certainly the preferred choice of the Defense Department. They loathed the idea of having to negotiate with a set of companies. A monopoly also made good sense from the customer's point of view. If subscriber A from city X wanted to call subscriber B from city Y, then he wanted a single company to assume responsibility for the success or failure of the call. Who would want to write seven complaining letters and who would want to hear the reply of the companies that any failure on the line was the fault of the other six companies?

In the 1930s the ideal solution appeared to be to acquiesce in AT&T's monopoly but to regulate their activities. How that view was slowly corroded will be the subject of Chapter 13.

Part III
The Modern Age Beckons

CHAPTER EIGHT

Great Advances

The Second World War gave an enormous push to all things technological but the main beneficiary was not communications: rather it was aircraft propulsion, rockets, radar, and of course the not-quite-so-peaceful side of nuclear engineering. Gigantic advances gave rise to gigantic confidence, best characterized by the American maxim: the difficult we do right away, the impossible takes a little longer.

Of greatest relevance to the advance of communications was the work on radar. After all, what radar does is to radiate in a narrow direction a pulse of electromagnetic waves which is reflected by the target and, from the time taken for the pulse to come back, the distance of the target from the radar set may be determined. The emphasis was on electromagnetic waves, the same waves that had been used for communications since the end of the previous century. Hence, the advances made during the war in producing electromagnetic waves could be easily translated to use those same waves for communications. The frequency range used by radar was in the microwave region (see Fig. 6.2 for the electromagnetic spectrum); hence, microwave links for communications were a simple continuation of the work done during the war.

A second advance was also due to radar. The devices which detected the reflected electromagnetic waves were made of semiconducting crystals. A semiconductor is neither a good nor a bad conductor of electricity. But if it conducts electricity at all, then it must contain some mobile electrons. If it contains mobile electrons, then one might be able to regulate the flow of those electrons. If one can regulate the flow, as in a vacuum tube, then it might be possible to produce an amplifier out of a semiconducting material. This was indeed the view in Bell Laboratories, the research laboratories of AT&T. They got going immediately after the war and by Christmas 1948 they produced the transistor, the most important electronic device of the twentieth century. Further development in this field brought the integrated circuit and the microprocessor, which led to the proliferation of personal computers.

A third advance was due to the development of rocket technology and to the imagination of Arthur C. Clarke, the science fiction writer. He envisaged a number of geostationary satellites (i.e. satellites which orbit at the same speed as the Earth rotates, hence appearing stationary to an earthling) which relay information from one end of the Earth to the other.

The fourth advance, the greatest theoretical advance, was the birth of information theory. This is concerned with the relationship between

Getting the Message: A History of Communications. Second Edition. Laszlo Solymar, Oxford University Press (2021).
© Laszlo Solymar. DOI: 10.1093/oso/9780198863007.003.0008

information and noise: to be precise, how noise will corrupt information. Its main tenet is just amazing. It took me a long time to absorb it. It maintains that if we do our best and we are not too greedy (i.e. we are happy not to exceed a certain rate when sending information), then it is possible to ensure that the information can be transferred *without a single error.*

The fifth advance is often referred to as the digital revolution but it should perhaps be called the digital evolution, because it came more in the form of a stream gradually expanding into a broad river rather than a torrent sweeping away everything. The idea was actually patented a little before the war but did not reach the market place until the 1960s. What's the good of it? It turns out that by putting information into digital form it can be processed much better. Not without a single error, that sounds a little utopian, but with present techniques error rates of less than one in a billion can be achieved routinely.

The sixth advance is optical communications, the seeds of which were planted some considerable time after the war. It brought together two elements: lasers and optical fibres. It enabled engineers to introduce light into a thin (about as thin as a human hair) strand of glass and guide it all the way to the other side of the Atlantic. The information to be transmitted, say the complete text and pictures of the *Encyclopaedia Britannica*, could be put on a light beam in London and the whole lot would reach New York before you could say 'Jack Robinson'.

The seventh advance was the emergence of mobile phones. Radio communication between any two points of the globe was, of course, available ever since short wave transmitters and receivers emerged. It was good indeed for broadcasting but not for individual calls. The motivation was always there; everyone wanted the chance to communicate cheaply and reliably with family and friends wherever they were but technology had to wait for many decades before it became affordable. Anyway, within a short time (see Chapter 14) mobile phones conquered the world.

The eighth advance is probably the most important of all. It enabled computers to talk to each other; a few in the beginning but soon burgeoned into a network, and later into a network of networks. They became fruitful and gave birth to electronic mail and to the Internet.

As the highways and the airways of the world get more and more crowded, contacts between human beings tend increasingly to move away from direct physical contact to seeing and hearing each other on screens. The exponential increase in the number of communication devices seems to go on unabated. We are spending greater and greater portions of our income on sending and acquiring information.

The administration and running of communication networks have also undergone major changes. They used to be thought of as a natural monopoly; i.e. the consumer got the best deal if it was run by a single organization, a company like AT&T in the US or various state agencies in most other parts of the world. This belief was undermined practically simultaneously around the middle of the 1980s in the two English-speaking countries, the US and the UK, by the divestiture of the Bell System in the former and by the privatization of British Telecom in the latter. All the world has been following that initiative ever since.

Finally, about genius in the post-war world: are there still great heroes in the field of communications? There are, although not in the sense that they have become household names. Nor would there be a clear consensus about who they are. It probably depends on where you live and whom you meet. I don't think there would be any dispute about the first three: they are William Shockley, John Bardeen, and Walter Brattain, the inventors of the transistor. They were before my time, anyway. As for the rest of my heroes, I met them all. Alec Reeves, who took out the first patent on pulse code modulation and who had tele-communications in his blood, was my boss while I worked at Standard Telecommunications Laboratories. All the others turned up at one time or another in Oxford. The man who made the greatest advance was Claude Shannon, who single-handedly founded information theory. Two of my heroes, Rudi Kompfner and John Pierce, may be seen in Fig. 9.10. They were both directors of Bell Research Laboratories at the time. Among many other feats, they were responsible for putting satel-lite communications into practice.[1] In this photograph Rudi is in his characteristic pose. He seems puzzled. But within a minute or so he is going to advance an explanation nobody ever thought of. A man whose impact on communications is still to come was Dennis Gabor. He invented holography.[2] It will be due to him if one day our moving images grow out of our boxes and appear in thin air in their full three-dimensional pomp.

I have ended up with seven heroes. A sign of the times is the American dominance. Four of the seven were born in the US (of the other three one was born in England, one in Austria, and one in Hungary) and even the Austrian-born did most of his work in the US. Will the American dominance continue? For the time being, yes. Where will technical leadership go next? Japan? China? Europe? I hope Europe, the great old civilization but my hopes have been declining in the past few years. At the moment the favourite looks like China where the Communist Party is busy building capitalism.

[1] For example, the travelling wave tube, the most ingenious of all microwave amplifiers, was invented by Rudi Kompfner and brought to perfection by John Pierce.
[2] He received the Nobel Prize for it in 1971.

Microwaves

The Quest for Higher Frequencies

Moving towards higher frequencies has two great advantages: more channels can be accommodated and it is easier to concentrate a beam in a narrow direction. A measure of the concentration achieved is the 'gain' which says how much more power is available at the receiver due to the directionality of the transmitting aerial.[1] This gain depends on the area of the aerial and on the wavelength. In fact, it is proportional to the total area of the aerial divided by the area of a square, each side of which is a wavelength (Fig. 9.1). For an aerial of 3 m diameter and a frequency of 1.7 GHz (corresponding to a wavelength of 17.6 cm) the gain turns out to be about 3000. Reducing the wavelength by a factor of 4, which would bring us near to some of the wavelengths used by satellite systems, the gain increases by a factor of 16. Thus, the saving in power resulting from the use of shorter wavelengths (higher frequencies) is really very large.

The move towards higher frequencies was bound to come. When it came, the jump in frequency turned out to be quite large. As early as 1931 the engineers of Standard Telephones and Cables, in collaboration with Laboratoires de Matériel Telephonique, set up communications between Dover and Calais at a frequency of 1.7 GHz. The aerial of 3 m diameter (Fig. 9.2) could produce such a concentrated beam that the output power needed was less than half a watt, not much if we consider that in most of our lamps we use bulbs consuming about 100 watts. Since the wavelength in these experiments was considerably smaller than those used previously, the new waves were initially called microrays and later microwaves.

Another company which got involved with high frequency waves in the early 1930s was Marconi's in Italy. They provided a link at a wavelength of 90 cm between the Vatican and the Pope's summer residence at Castel Gandolfo 15 km away. The story goes that there was some disturbance in the communications every day around the same time. The cause was discovered to be a steamroller moving across the path of the microwave beam. The disturbance came from the waves reflected by the steamroller. But then the question could be asked: if the presence of a steamroller can be determined by reflection of microwaves, why not the presence and positions of ships or aircraft? The incident is said to have helped to focus attention on the possibilities of using microwaves

[1] Microwave transmission and reception is usually with parabolic dishes which have been a common (perhaps too common) sight since the advent of satellite television channels.

Getting the Message: A History of Communications. Second Edition. Laszlo Solymar, Oxford University Press (2021).
© Laszlo Solymar. DOI: 10.1093/oso/9780198863007.003.0009

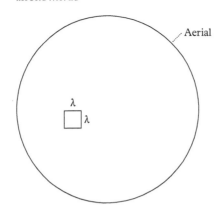

Fig. 9.1 Schematic represen-tation of a dish aerial and an area each side of which is equal to the wavelength.

to determine position. This is not to say that the invention of radar was in any way serendipitous. As noted in Chapter 6 Hulsmeyer in 1904 did already observe reflections from ships. Marconi himself had been well aware both of the military significance of radar and of the technical possibilities well before the incident with the steamroller. In fact, he mentioned the possibility of determining position with the aid of radio waves in a lecture he gave in New York in 1922.

It was certainly Marconi who was responsible for starting off radar development in Italy. By 1934 he could demonstrate his equipment to Mussolini and to the military top brass. Had he not been in poor health and had he not died in 1937, Italian radar would have been at a much higher state of development. As it happened at the Battle of Cape Matapan in 1941 the Italian Navy, in contrast to the British, was still not equipped with radar. The British ships were capable of opening fire in the dark from a distance of 3 km and quickly sank three Italian cruisers and two destroyers.

In the 1930s all the radar work in the various countries was done at frequencies well below 1 GHz (wavelengths well above 30 cm). It was important to move towards higher frequencies (shorter wavelengths) because for accurate determination of position, narrow beams are needed and narrow beams can only be achieved at higher frequencies. Nobody knew how to achieve these higher frequencies. It was not simply a question of putting in sufficient effort. An entirely new departure was needed, beyond the capabilities of any of the Post Office Departments in the UK. A number of academics were recruited from various univer-sities. Their approach to research did occasionally clash with civil ser-vice orthodoxy. A prominent member of the group, H. W. B. Skinner, was once warned about arriving late to work and J. Atkinson was sent off to get his hair cut. The initial difficulties are well illustrated by Skinner's poem:

This parabolic reflector, three metres in diameter, was set up on the cliffs at St. Margaret's Bay, Dover, for the world's first public demonstration of microwave transmission on April 3rd, 1931.

Fig. 9.2 This portable reflector, 3 m in diameter, was set up on the cliffs at St Margaret's Bay, Dover, for the world's first public demonstration of microwave transmission on 3 April 1931.

We worked all through those early months, fighting,
first to be allowed to work at all,
and then for space and light and heat to work in;
for we were thought fit but for the next war,
or next year's war at best, and next year
was like a mirage in the desert sky,
when June had drained the last drops of complacency,
and France had written finis to that phase.
And so alone,
we, fighting every inch of the way,
against those ingrained elephants of inertia,
against the prejudice and the pride
of self-established, self-supporting systems,
we fought (through forests, thick with self-satisfaction)
to shorter electromagnetic wavelengths.

The first breakthrough, the most important invention, was that of the cavity magnetron in 1940 by Boot and Randall of Birmingham University.[2] Soon it could produce peak powers up to 100 kW at a wavelength of 3 cm. Another significant invention made in Britain was the so-called travelling wave tube, which made possible the wide-band amplification of microwave signals. The inventor was a Viennese architect, Rudi Kompfner, who emigrated to Britain in 1934.

Microwave Links

By the end of the war microwave techniques were well worked out. At the same time, at least in the US, the demand for communications grew so sharply that it would have been difficult to satisfy it by using overhead lines and cables. The way to do it with microwaves was to build a chain of towers which looked upon each other (Fig. 9.3). The idea was the same as with beacons or the mechanical telegraph. Each tower was a relay station. The main advantage was the large number of channels available, but there was another advantage too. To lay a cable one needs to negotiate with those owning the land, and landowners are not famous for their altruism. It costs money. To build a tower also costs money, but much less. It was the possibility of such a low-cost incursion into communications that caused the fall of the Bell System in the end. That will come later in Chapter 13.

The first microwave link was built between New York and Boston at the end of the 1940s, relying on the band 3.7 to 4.2 GHz. The distance of 220 miles was covered by five relay stations with an average distance between towers of about 36 miles.[3] A later system working in the 6.6-GHz region provided 26,000 channels in total. The modulation system chosen was frequency modulation, mainly because the corresponding amplifier specifications were less strict. With improved amplifiers they

[2] Little was it expected at the time that the magnetron's main application would eventually be for purposes unconnected with war. It is the device which provides the heating in all our microwave cookers.

[3] Not quite as efficient as the early warning system of Leo the Mathematician, who managed to cover the 450 miles from Constantinople to Loulon with relays 45 miles apart.

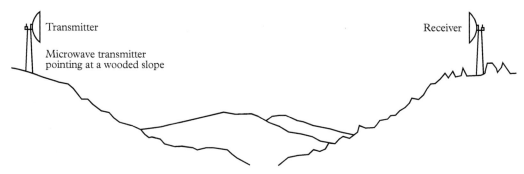

Transmitter

Microwave transmitter
pointing at a wooded slope

Receiver

Fig. 9.3 Schematic representation of a microwave link between a transmitting and a receiving aerial.

could later change to single-sideband amplitude modulation, thereby increasing the number of channels to 42,000.

Many other microwave links followed. By 1967 the whole of the United States was densely covered by this new means of communications, as may be seen in Fig. 9.4. In the UK the first link was built between Manchester and Kirk o'Shotts in Scotland. It opened for service in 1952. In the UK expansion was quite rapid. Soon microwave links covered the whole country, including a link to Northern Ireland. Its centre was the Post Office Tower in London. Figure 9.5 shows the microwave antennas on top. Microwave links crossed the Channel in 1959. The equipment was made by Standard Telephone and Cables in Britain and by the Laboratoire de Matériel Telephonique in France, repeating their feat of 28 years earlier. Thanks to similar efforts in Western Europe television images of the Rome Olympics in 1960 could be seen live in England.

Finally, it may be worth mentioning a particularly arduous exercise of building microwave links in the Caribbean Islands, carried out by the engineers of Cable and Wireless between 1971 and 1977. Since many of the towers could be erected on mountain tops, fewer stations sufficed, but the mountain peaks were accessible only by a combination of high-tech helicopters and low-tech donkeys.

I learned of another interesting case from a French friend of mine whose job was to set up microwave links in Gabon. The mountain most suitable for a relay tower was unfortunately the dwelling place of a god worshipped locally who, according to the local population, would not look kindly at any disturbance. The solution found eventually was to have a feast for everyone where sacrifices were offered to the god. The sacrifices were accepted, the god was placated, and the building work could start and finish without any divine interference.

Waveguides

The advantage of microwave links was that lots of channels could be transmitted without needing to lay cables in the ground. Does that

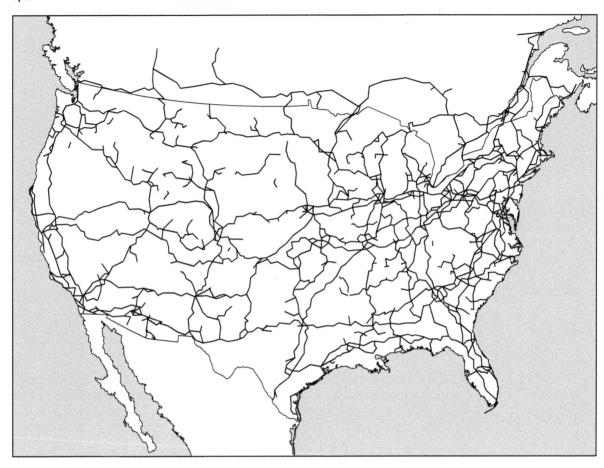

Fig. 9.4 Microwave link network in the US in 1966.

mean that guiding structures (microwave cables are usually referred to as waveguides or, more generally, as guiding structures) were entirely unsuitable for microwave communications? No, they were not. At one time the prospects looked exciting.

Waveguides are made of hollow metal tubes which may have rectangular or circular cross-sections. The electromagnetic waves carry the information inside the tubes very much like water. Information flows in at one end and comes out at the other end. An alternative view is to look at wave propagation as a result of a series of reflections by the walls, as shown in Fig. 9.6 for a rectangular waveguide.

Unfortunately waveguides, like any other guiding structures, have attenuation which depends on the size of the waveguide and on the electric field configuration within them. Attenuation usually increases with frequency, so that if we want to accommodate lots of channels and employ higher and higher frequencies, the waveguides become less and less suitable for information transmission. There is, however, an interesting exception, discovered theoretically in the 1930s by Sally Mead of Bell

Fig. 9.5 The Post Office Tower in London.

Laboratories. For one particular electric field configuration the attenuation of a circular waveguide declines with frequency. Taking a fairly large waveguide diameter of 7.5 cm and a frequency of 75 GHz (a wavelength of 4 mm), the attenuation is no more than 0.23 db/km.[4] Thus, the attenuation is no problem. What about the number of telephone channels available? Assuming frequencies between 30 and 100 GHz the number of available channels with the most economic single-sideband modulation comes to 1.75 million, or to about 100,000 if PCM (pulse code modulation; see Chapter 11) is used. No wonder that communications engineers got excited about it. But why use waveguides which need to be buried in somebody else's land, and pay rents? Why abandon the chain of towers? The answer is that atmospheric absorption is unduly high for such high frequencies so free space propagation is out of the question. On the other hand, with the prospect of 100,000 channels it will still be economic to pay rent and undertake all the expenses of manufacturing and laying the waveguides.

Thus, the motivation to go ahead was quite strong. Research started a few years after the end of the war, more or less simultaneously in the US and in Britain. The British effort was concentrated at Standard Telecommunications Laboratories (the principal research laboratory of International Telephone and Telegraph), at University College, London and at the Post Office Research Station. The American research was mainly conducted at Bell Laboratories. By the middle 1960s most of the technical problems had been solved. The only thing needed was customers. Unfortunately, there were no two cities anywhere in the world which would have cried out for 100,000 brand new telephone channels. And besides, the system would have been quite expensive to install. AT&T's experimental line cost $100,000 per mile.

AT&T hoped to stimulate demand with its Picturephone, which was the sensation of the 1964 New York World Fair. It was introduced in service in Pittsburgh in the summer of 1970. Had it caught on it would have gobbled up bandwidth, one picture channel being equal to one hundred telephone channels, but as it happened the public wanted to be heard but not seen.[5] The British Post Office and AT&T had no other option but to wait patiently for demand to pick up but before this happened, suddenly, a competitor to the microwave waveguide appeared in the form of the optical fibre (see Chapter 12). It was enormously bad luck for those working on microwaves. It often happens in the history of technology that the solution envisaged for a particular problem turns out to be

[4] Attenuation is usually measured in db/km (pronounced decibel per kilometre), which is a rather complicated logarithmic measure. For example, 10, 3, and 0.2 db stand for a decline in power by factors of 10, 2, and 0.045 (4.5%), respectively.

[5] A year later there were only 33 Picturephones operating in the whole city.

Fig. 9.6 A wave may propagate by subsequent reflections from a waveguide wall.

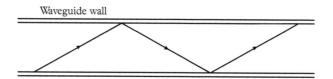

Waveguide wall

deficient in one or another aspect and the designers have to go back to the drawing board. As far as I know long-distance communications by microwave waveguide is the only case when a technically excellent solution never saw the light of the day because, before it could be put into practice, a series of new inventions made possible a cheaper, alternative solution. As someone who worked on the project for a couple of years I mourned its demise. The new solution, better and cheaper, was provided by optical fibres (see Chapter 12).

Satellite Fundamentals

Satellite is a word that has been in use ever since antiquity. According to the Oxford English Dictionary it means "An attendant upon a person of importance forming part of his retinue and employed to execute his wishes". A more modern dictionary would define it as: "an artificial body put in orbit around the earth or another planet." How can it be used for communications? The idea is simple. If a body can orbit the earth then it must be possible to send a signal to it and if that body contains a suitable electronic device then it could resend the signal to some other part of the earth establishing a communications link. The idea is simple, the realisation is not so simple.

First of all, can we have a satellite that obligingly hovers above the Earth and stays there for ever, or at least for five years or longer? To explain that, we need a little bit of science. If you whirl a small object fixed to the end of a rope, you feel a force upon your hand that tries to tear the object away. The force is perpendicular to the path the object describes. It is called the centrifugal force. If you want that object to move round and round without the rope that is also possible. The centrifugal force on the object can be balanced by the gravitational force of the Earth. The derivation is not too difficult. It is given in Appendix 2.

What kind of satellite do we need? We need one that moves fast but remains at the same place. With remarkable foresight this is exactly what the Mad Hatter suggested: "It takes all the running you can do, to keep in the same place", he said. Well, it has been done. The satellite that can do that is called a geostationary satellite. It moves fast with a speed equal to the rotational speed of the Earth; hence, for us earthlings it appears as if it were stationary. That's a great thing, for it is always there and we can rely on it. We can send up to the satellite a microwave beam carrying lots of information which it reradiates to another part of the Earth with which we want to communicate. We must add that this miraculous achievement of moving but staying in the same place can occur only in an orbit around the equator and only at an altitude of 35,786 km above the Earth's surface. This is quite advantageous because from that height one can see a large part of the Earth, up to 81°

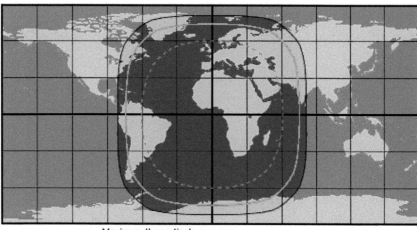

— Maximum theoretical coverage
— Imaging and telecommunications coverage
- - - Meteosat images within this area used quantitatively

Fig. 9.7

of latitude. The coverage by one of these geostationay satellites is shown in Fig. 9.7. In principle three satellites would be enough but reception would be dodgy at the edges, so in practice five or six satellites give a much better coverage. Note, however, that there is not much business at the poles; hence, some companies have decided not to cover those areas.

There are, of course, other orbits too where the two forces just balance each other. For each altitude there is a velocity that makes that possible. In orbits below the geostationary the satellites move faster. In orbits above the geostationary the satellites move more slowly. In the geostationary orbit they are just stationary. It is not necessary for the satellite either to move parallel with the equator or to have a circular orbit. There could be an orbital inclination as shown in Fig. 9.8a, and examples of elliptic orbits are shown in Fig. 9.8b, respectively. Geostationary orbits are nice and useful. So is there any incentive to use different orbits? There is a very strong one. The problem with geostationary satellites is that they are too far away. Well, as long as they keep to their orbit, why does altitude matter? Is there perhaps a danger that if they are too far away they might somehow escape? No, we can rely on gravitation.[6] So what's wrong? The culprit is the finite velocity of the electromagnetic waves. It is 300,000 km/s. To the geostationary satellite and back the distance is 75,000 km, corresponding to a quarter of a second. Does it matter? One would think that anybody is willing to wait a quarter of a second for a response. Oddly enough it is not true. Such a delay is quite noticeable, mostly in television interviews. The delayed reaction of the interviewee is clearly visible on the screen. In an ideal communication system such delay is undesirable. But there is a second problem too. The satellite being too far away means that a lot more power is needed to communicate with it. More power means more

[6] Being far away is no problem for stability. The Moon is 384,000 km away from the Earth and shows no signs of wanting to severe its connection. Or Pluto is another example. It is 5.8 billion km away from the Sun and it is still strongly held. The effect of distance is that it takes longer to circle the Sun. To be precise, for Pluto it takes 248 years.

(a)

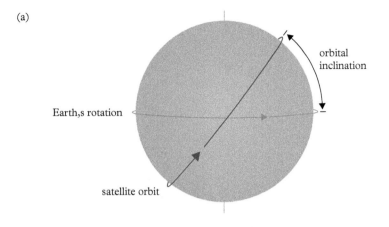

(b)

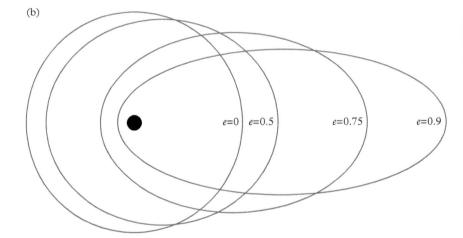

Fig. 9.8 Satellite orbits. (a) Black, geostationary orbit; red, circular inclined orbit. Redrawn. Source: (b) Elliptic orbits. The eccentricity e is a measure of the ellipticity. NASA Earth Observatory.

batteries, more batteries means more weight. So if we want to use a light handset, we should not use geostationary satellites.

So the criterion of a good communication system nowadays is the negligible delay of the signal plus a light handset, plus coverage of the Earth. The first and second aims can be achieved by choosing a lower Earth orbit, LEO in the official jargon. To achieve the third aim is more difficult. The company Iridium SSC rose to the challenge in the late 1980s. Iridium was a properly chosen name for the exercise. Iridium is a metal with atomic number 77. According to the early design 77 satellites were needed (see the orbits in Fig. 9.9) to provide perfect coverage from any point of Earth to any other point.

The Beginning of Satellite Communications

Microwave links can span continents but cannot offer communications across the oceans. In principle it would have been possible to lay a microwave waveguide under the ocean but it did not even become a practical proposition on land, let alone under the sea. Was there a third microwave alternative? Yes, it was proposed in 1945, not by an expert on telecommunication techniques but by a renowned writer of science fiction, Arthur C. Clarke. This is hardly surprising considering that satellite communications are closely related to the idea of space travel of which those boys and girls who read Jules Verne and H. G. Wells knew. But Clarke was more than a science fiction writer. He was perfectly aware of all the principles involved. His paper published in *Wireless World* outlined all the main features of communicating via manned satellites. The title of the paper was 'Extraterrestrial Relays'. As Clarke wrote later, 'the paper met with monumental indifference'. He did his best though to promote the idea the only way he could. There were lots of manned communications systems in his novels *Prelude to Space* (1950) and *The Exploration of Space* (1952). The latter book sold about a million copies, so there must have been a number of men and women on Earth well aware of the potential of communicating through the skies. But it is a different thing to read science fiction and to think about practical realization.

First, there was an obvious need for a sufficiently powerful rocket to put the satellite into orbit. In the US this was established by a team led by Dr Wernher von Braun, a German engineer who, after the end of the Second World War, suddenly changed his interest from blasting London Town to smithereens to the peaceful joys of space travel. By 1954 the idea of space travel was already 'in the air'. To bring it down to Earth and back into the sky an influential man was needed, willing to look at all the practical details and mobilize resources for a successful demonstration.

Fig. 9.9 Iridium's original 11 satellites in 7 polar orbits of the Earth. By the time the project came to realization in the 1990s, it turned out that 66 satellites were sufficient. Six orbital planes were chosen 30° apart, with 11 satellites in each orbit. The reason for needing so many satellites was the low orbit at 781 km. The time delay was practically abolished but the price paid was having many satellites. It needed many satellites in the same orbit because at that altitude the orbital period was only 100 minutes in contrast to one day for geostationary satellites. That means that any of the satellites is available for only about 15 minutes and then the next one takes over but, fortunately, the next one is always there.

Fig. 9.10 Antenna One, named 'Arthur', at Goonhilly Earth Station. Source: Carcharoth.

[7] For satellite communications to work it was necessary to have a powerful amplifier. Such an amplifier, the travelling wave tube, was invented by Rudi Kompfner, a Viennese architect who emigrated to Britain in the 1930s. At the beginning of the Second World War he was interned as an enemy alien. By that time he was interested in electronics. He had a deep understanding of physics that enabled him to think of radically new electronic devices. His ability recognized, Kompfner was catapulted into a research laboratory run by the Admiralty. A decade later he joined Bell Telephone Laboratories in New York, where he worked closely with John Pierce.

It is worth telling a story he told me many years ago. One night there was a terrible hullaballoo in the neighbouring barrack occupied by Japanese aliens. The noise was such that it was impossible to sleep. Next day, a delegation was sent to the camp commander complaining about the noise. 'Very sorry', said the commander, 'I had to give them permission to celebrate the fall of Singapore.'

Such a man was John Pierce of Bell Telephone Laboratories. In Fig. 9.10 he is shown in the company of his colleague Rudi Kompfner: the two objects on the table are travelling wave tubes.[7] Pierce's first paper on satellite communications was presented in October 1954. His second paper put the subject on the agenda for commercial exploitation and, of course, the Defense Department soon recognized that they had acquired a new toy to play with which would very likely solve their communications problems for ever and ever.

How did it all start? The first satellite was launched in October 1957 by the Soviet Union. They called it a Sputnik (meaning fellow-traveller), introducing thereby a new word in the English language. They beat the US by three months. The first satellite set up for communications purposes was called appropriately Echo. It was a passive one, relying on the bouncing off of signals from its surface. The signal coming from America could be received at Goonhilly Earth Station in Britain where

Fig. 9.11 John Pierce and Rudi Kompfner in discussion.

8 'Arthur', in honour of Arthur C. Clarke, is a massive steerable antenna weighing 1,100 tonnes, able to track at a speed of 120° per minute with an accuracy of better than one hundredth of a degree.

there was a big enough antenna (see Fig. 9.11) to capture the signal.[8] This was the way we were able to obtain television from the US, and how the Americans could watch Olympic Games held in Europe. For today's world satellites play an even more important role. They make possible telephone calls and data transfer between any two points on Earth.

Fig. 9.12 Telstar, the first active satellite.

The first active satellite, provided by a transponder, was Telstar (Fig. 9.12), which went into orbit in 1962. It established for the first time live television transmission between the new and the old world. In her 1962 Christmas broadcast to the Commonwealth Queen Elizabeth referred to Telstar as 'the invisible focus of a million eyes'. The first geostationary satellite, Early Bird, was launched in 1965. It provided one television channel or 240 two-way telephone channels, not the kind of numbers we are used to nowadays but at the time it effectively doubled the transatlantic telephone capacity.

Thanks to geostationary satellites it became technically possible to contact London from the Amazonian jungles or from a construction site in Africa. The equipment needed was transportable (Fig. 9.13) but not a simple handset.

Let us now return to the fate of Iridium SSC, the company that was the first to set up a sophisticated satellite constellation. The satellites were deployed between 1997 and 2002 using some American rocket-launching sites but mainly Russian and Chinese ones, presumably because they were cheaper. The whole exercise cost some US$5 billion. Full global coverage was achieved by 2002. It worked as it was supposed to. It worked, in fact, very well. Unfortunately, there was no revenue

Fig. 9.13 An early mobile phone. Source: GloCall Satellite Services.

Table 9.1 Comparisons between three satellite communications providers

	Globalstar	Iridium SSC	ICO
Number of satellites in orbit	48	66	10
Number of planes	8	6	2
Orbit altitude (km)	1,414	781	10,350
Output power (watt)	1,000	1,400	2,500
Launched	1999	1998	2000
Data throughput (kb/s)	7.2	2.4	2.4
Intersatellite communications	No	Yes	No

before the whole thing was set up and even when revenue started to come in, it was insufficient to service the accumulated debts. Iridium SSC was forced to declare bankruptcy. It was the largest bankruptcy in the US at the time. Fortunately for the company, there is a type of bankruptcy in the US that makes resurrection possible. It is called a Chapter 11 bankruptcy. All the debts can be restructured and the company is allowed to continue operation. That's what happened. The company rose from its ashes (like a phoenix) under the name of Iridium NEXT but that is another story (see Chapters 14 and 17).

Iridium was the first but not the only company that set up a satellite constellation. It would be too boring to mention all of them. However, in Table 9.1, I provide some information about two further companies, Globalstar and ICO, and compare them with Iridium SSC.

10

Devices Go Solid State

The Background

Solid state physics, as the name implies, is concerned with the physical properties of solid bodies, in particular what happens to them when they are exposed to external influences, e.g. heated, voltage applied across them, or light shone upon them. This was, of course, a science that was already flourishing in the nineteenth century but with serious limitations. The theories available at the time could explain only a fraction of the experimentally found phenomena. Entirely new departures were needed.

The new physics advanced in two main directions. The physics of big things (relativity, cosmology, etc.) and the physics of small things (how atoms behave, what's inside them), which later became known as quantum physics. Solid bodies consist of lots of little atoms piled upon each other. Therefore, theories of the solid state had to wait until the problem of individual atoms was properly solved. This occurred in the late 1920s mainly as a result of the work of Erwin Schrödinger, Werner Heisenberg, Max Born, and Paul Dirac. They were all theoretical physicists.[1] Their theories were entirely incomprehensible to all but a select band of their colleagues. They threw old physics out of the window, abandoned common sense, and replaced them all with a rather strange mathematical apparatus. Built on this foundation, theories of solids, of metals, of insulators (which do not conduct electricity), and of semiconductors (which conduct electricity a bit) were developed. One of the triumphs of the new theory was that it could explain the apparent presence of positively charged mobile particles, despite the fact that the only mobile particles actually in the solid were negatively charged electrons. Those positive particles acquired the prosaic name 'holes',[2] which shows that name counts a lot when it comes to fame. Whoever heard of holes? Only the specialists. A Greek-sounding name with a 'tron' at the end would have served them better.

The new theories had considerable success in explaining properties of solids (e.g. the specific heat of metals) where the predictions of classical theories were not even in the right ballpark. But detailed comparison between theory and experiment was not possible for two good reasons: firstly, the theories were too rough, and secondly, the technology available was not suitable for producing materials of sufficient purity.

[1] Theoretical physics was a unique European invention. No other civilizations took the trouble to look into its mysteries.

[2] There was some rationale behind the choice. Missing electrons (in other words, holes in the electron population) may behave as positive particles.

Getting the Message: A History of Communications. Second Edition. Laszlo Solymar, Oxford University Press (2021).
© Laszlo Solymar. DOI: 10.1093/oso/9780198863007.003.0010

What was the motivation for all this work on the properties of the solid state? Curiosity. Nobody would have denied, of course, that practical applications would come one day but that was not why they did it.

The Invention of the Transistor

By the end of the Second World War some people strongly believed that the time for solid state amplifiers had arrived. During the war lots of advances had been made in producing purer crystals for microwave rectifiers, so the technological problems no longer looked so daunting. And the motivation was there too. Vacuum tubes were examined and found wanting. They were clumsy, expensive, and unreliable. And progress wasn't very fast either. It had been a good thirty years since the invention of the triode amplifier, and a telephone cable under the Atlantic was still only a dream. An amplifier that does not rely on evacuated tubes was badly needed. At such times the right approach is to follow a hunch and to be generous with resources. This is what Mervin Kelly, the director of research at Bell Laboratories, did just after the end of the Second World War. He decided that the operational principles of the alternative device lay hidden somewhere in the electronic properties of solid materials, and he set up a group under the leadership of William Shockley to look into the matter. Among members of Shockley's team were John Bardeen and Walter Brattain.

When Shockley's team started their investigations, a fair amount was already known about the properties of electrons and holes. It was known, for example, that semiconductor materials could be made n-type, in which electrons dominated, or p-type, in which holes were the dominant population. It was also known that n-type and p-type materials, when joined together, acted as rectifiers in a similar way to Fleming's diode. The aim of the group was twofold: to understand the properties of semiconductors in general and to try to realize an amplifier which had been proposed as early as 1925 by Julius Lilienfeld, a professor at Leipzig, and which had stubbornly refused to work as it was supposed to. The idea was very simple. A voltage applied across a piece of semiconductor was supposed to affect the current flowing perpendicular to it.

The experiments conducted at Bell Laboratories provided the same negative result as all the previous experiments. The effect just wasn't there. But there was a difference. They knew more about semiconductors than earlier experimenters so they were in a position to find the reason why the expected effect was absent. John Bardeen showed theoretically that the surface between the semiconductor and the metal electrode (by which the voltage was applied) could trap the electrons, and that could account for the failure of the experiment. So they looked at the surface in a systematic manner, relying on the experimental genius of Walter

Brattain. There was a lot to learn about the surface but before they learned the answers to all the questions they asked, they stumbled upon an amplifier. It was not the amplification effect they looked for, but nonetheless it was amplification and, after all, what they wanted was an amplifier, so everyone was very happy. When they were sure that the amplifying effect was reproducible they presented the new device to the management of Bell Laboratories. This occurred just before Christmas in 1947. All those present were allowed to say a few words into a microphone and, lo and behold, an amplified sound came out of the loudspeaker. The occasion was also used for the baptism ceremony. John Pierce, whom we met in Chapter 9 on microwaves, coined the word 'transistor'. It sounded OK. It broadcast immediately the arrival of a new and important electronic component, which by that time included resistors, capacitors, inductors, thermistors, and varistors. The 'trans' at the beginning referred to the working of the device as a transconductance.

In the nineteenth century the accidental discovery of an amplification effect would not have unduly worried the discoverers. The main thing was that it worked. Other things were of less importance. In the atmosphere of Bell Laboratories in 1947 the primary goal was to find the physical mechanism. Shockley, not having much to do with the accidental discovery, decided that it was his duty to provide an explanation. He looked theoretically at the dynamics of electrons and holes in thin layers of different types of semiconductors joined together. He thought that was an appropriate model of the device and he got cracking. He worked through New Year's Eve producing 19 pages in his notebook. On 23 January 1948 the concept of a three-layer amplifier was born. He could even work out how the amplification depended on the properties of each layer. Was this the explanation they had all been looking for? Nobody was certain, but it did not really matter. Even if it was the wrong explanation for the existing device, it was surely worth building a new one based on Shockley's recipe. The device was built and worked more or less as expected. It took two more years of painstaking work for the transistor to reach maturity, in the sense that all the major effects had been understood and manufacturing techniques had been developed. By this time the device discovered by Bardeen and Brattain was known as the point-contact transistor. Accurate reproduction of its properties was difficult; it went out of production by the end of the 1950s. Shockley's device, which came to be referred to as the junction transistor, went from strength to strength. Today we call it a bipolar transistor (because charge carriers of both polarities play important roles), or just a transistor without any prefix. The three inventors, Bardeen,[3] Brattain, and Shockley, received the Nobel Prize for their invention in 1956.[4]

AT&T could have reserved for themselves the right of manufacturing transistors for the lifetime of the patents—25 years. Some of that extra

[3] Bardeen received a second Nobel Prize in 1972 for his contributions to the theory of superconductivity.

[4] When assigning credit to those who made the invention of the transistor in the United States possible, one should not forget to mention the pivotal role played by Adolf Hitler. There is no doubt that some American scientists had already established an international reputation in the interwar period. Between the years 1919 and 1939 there were five Americans among the 24 Nobel Prize winners in physics. Not a bad start but hardly an indication of America leading the world. The new scientific ethos was, to a large extent, created in the 1930s by émigré scientists crossing the Atlantic under Hitler's persuasive influence. The direct effect on those working on the transistor was perhaps not that great (although Bardeen was a graduate student of one of the émigrés, Eugene Wigner) but the whole atmosphere had changed.

revenue would have been used to provide better services; the rest would have gone into the shareholders' pockets. Would such a policy have benefited AT&T? Well, it would have made it richer and more powerful, but perhaps too powerful. In the changed post-war world the image of a selfish giant was bad for public relations, and besides, it might have induced the US Department of Justice to look with less benevolent eyes at the near-monopoly of AT&T in the telecom business. So there were strong incentives to conduct a less selfish policy and that's what AT&T did. They offered manufacturing licences to all and sundry for the very moderate sum of $25,000. Within a few years more than two dozen American companies bought the licence, and so did a Japanese company which rose to fame later under the name of Sony. Interestingly, their application for a license was held up by the Japanese Government for a full year because they were anxious about such a large amount of foreign currency leaving the country.

There is no doubt that enlightened self-interest played a major role in making the transistor available for everybody on equal terms. But perhaps one should not dismiss the possibility of an element of genuine altruism. Here is a company, in terms of market value the largest in the world, which is proud to have had the foresight to move into semiconductor research, is proud to have produced the most important device of the century, and wants the whole world to benefit from the invention.

Let me add here that the original quest for the device proposed by Lilienfeld went on. It was proven by Shockley and Gerald Pearson in 1948 that the effect did exist, but a lot of further development was needed. The first commercially available device, called the Technitron, was produced in France in 1958 by Stanislaus Teszner. In the US, it became available in 1960. Owing to its resemblance to the transistor but in acknowledgement of the fact that the principles of its operation were different, it became known as a field effect transistor, abbreviated to FET. There was a race for dominance between the bipolar transistor and the field effect transistor which has apparently been won by the latter. The number of field effect transistors manufactured in 1998 was 1,500,000,000,000,000, which is about 300 000 for every man, woman, and child on Earth. What is the number today? Nobody knows. It is no longer of any interest.

The Transistor Goes West

The first transistors were made of a semiconductor called germanium. It had the disadvantage that it could not work at high temperatures. Silicon, another semiconductor, was much more suitable for higher temperatures (which inevitably occur in many devices) but its technology was more demanding. Bell Laboratories were, of course, involved in trying to make silicon technology work. By this time Shockley was not there. He had a desk job at the Pentagon as an eminent civilian scientist but he was not too happy with it. He would have liked to return to industry or, rather, to strike out on his own. His original aim was to set up a company which could produce silicon transistors and could, of course, further advance the art. He found a financial backer in the person of Arnold Beckman. He recruited a number of bright young men, partly from some of the smaller companies and partly by scouring the top universities for their best PhD students. Among the young men recruited were Robert Noyce and Gordon Moore, who were to play significant roles in the revolution.

The next question was where the new company should be located. Shockley decided to go back to where he grew up, to the San Francisco Bay area. That's where the Shockley Semiconductor Laboratory took its first steps in 1956 in establishing silicon technology. The zenith was reached later in the year when Shockley received the Nobel Prize. When the news arrived, the staff were only too happy to celebrate their boss (Fig. 10.1).

Later times were less happy. Shockley turned out to be a poor businessman and an inefficient manager. A lot of money went into the business but very little came out of it. The staff were unhappy too because Shockley kept on changing his priorities. In September 1957 eight of his top men resigned (Shockley called them the 'traitorous eight'). With finances from the Fairchild Camera and Instrument Company they founded Fairchild Semiconductors, one mile down the road from their former employer.

Shockley soldiered on for a while but his company never made a profit. He moved into academic life in 1963. The men he had hired made history. They became fruitful, multiplied, and replenished the whole neighbouring area, and while doing so they turned it into the biggest wealth-generating machine in the world, known as Silicon Valley. Shockley had the chance, he was surely the anointed man, but he could never enter the promised land because of his failings in the art of management. The analogy may not be very close but it is there. Shockley was said to be the Moses of Silicon Valley.

Computers

It is well beyond the scope of the present book to explain the operation of computers from beginning to end. A brief introduction is, however, necessary because the interaction between computers and communicatons has always been strong. By now it is impossible to say where the computer ends and communications equipment starts. This is now called, a bit pretentiously, the communications–computing symbiosis. I shall come back to it in Chapter 16 where I shall discuss computer networks.

Fig. 10.1 Senior staff of the Shockley Semiconductor Laboratory toast their boss at a lunch the day after the announcement of his Nobel Prize in 1956. Gordon Moore is sitting at the far left; Robert Noyce is standing fourth from left. William Shockley is seated at the head of the table.

[5] One of the hazards was the glow of the tubes which, apparently, attracted moths, causing frequent short circuits. To clear away the moths was called 'debugging', a term used ever since for finding out what's wrong with a program.

Computers are, of course, terribly complicated, awe-inspring things, but they are put together from extremely simple components. Essentially they consist of an assembly of switches to which peripheral devices like screens and printers are added. A computer can do all the arithmetical operations (and any higher mathematics is reduced to that eventually) with the aid of so-called logic gates, performing operations in logic.

Early electronic computers used vacuum tubes as switches, which had a rather limited life and consumed an inordinate amount of power. Colossus, the first operational computer, was built in 1943 by the British Post Office for codebreaking. It had 2,500 tubes and consumed 4.5 kW of power. ENIAC (standing for Electronic Numerical Integrator and Computer), completed somewhat later in the US, was even more colossal with 18,000 tubes and a power consumption of 140 kW. A limit in the number of tubes was soon reached due to the constant need to replace the faulty ones.[5]

With the advent of the transistor reliability went up, power consumption went down, and the size of the computer decreased considerably. Very soon computers employed tens of thousands of transistors and hundreds of thousands of other kinds of components (resistors, capacitors, diodes). Considering that a transistor had three terminals and the other components had two terminals each, there were lots of terminals to be connected to each other. One can imagine the time and effort needed to do all that soldering. And, of course, the chances of not making a single mistake in all that interconnection jungle were rather small. The 'tyranny of numbers', as someone in Bell Laboratories called the problem, had to be tackled.

Integrated Circuits

A solution to the interconnection problem was outlined by G. W. A. Dummer of the Royal Radar Establishment as early as 1952. He wrote:

With the advent of the transistor and the work in semiconductors generally, it seems now possible to envisage electronics equipment in a solid block with no connecting wires. The block may consist of layers of insulating, conducting, rectifying and amplifying materials, the electrical functions being connected directly by cutting out areas of the various layers.

The idea was there but it was rather vague. Attempts to put the ideas into practice were pursued in the UK at a rather leisurely pace. The British Empire was in decline. There was no great pressure from the military for immediate results. The situation was entirely different in the US. The Cold War was in full swing. The armed services had plenty of resources available and they desperately needed small, rugged, reliable electronic devices to be used in missile guidance systems. Each of the services had its own pet project.

The most forward-looking one was the Air Force. They wanted a clean start, relying on an entirely new approach called 'molecular electronics'. The idea was that layers of molecules laid upon each other would be able to perform all the necessary functions. Unfortunately, there were neither experimental nor theoretical foundations to build on. The project seemed unlikely to succeed. The Army initiated a 'modular' approach, trying to make all the components the same size and shape so that they could then be snapped together without the need for soldering. That did not seem to work either. The Navy was in favour of 'thin films', that is, printing components on a ceramic base. Some advances were made but the chances of a breakthrough were remote.

That was the situation in the summer of 1958 when Jack Kilby appeared on the scene. He joined a small firm, Texas Instruments, just before the staff went on their annual holidays. Had that happened in Europe, the management would have kindly let him enjoy an undeserved rest. America being America, Jack Kilby had to stay in the laboratory more or less on his own. He was asked to work on the 'modular' solution financed by the Army. He knew it was the wrong approach. He spent the next two weeks trying to think of a better solution. He came up with the idea that silicon, the semiconductor employed by Texas Instruments for producing transistors, could also be used for producing resistors and capacitors. He argued that once all the components were manufactured simultaneously by the same technique on the same substrate (i.e. they are integrated), then their interconnections by a conducting material, such as gold, laid down on the insulating surface would be fairly easy.

Was this a great idea? It was the greatest since the invention of the transistor. Was it obvious? No, it was not obvious at all. The production of resistors and capacitors was an established technique. Nobody in his right mind would use silicon, an unsuitable material, for either of them. However, Jack Kilby had faith in his proposal and managed to persuade his boss to try it. Less than two months later, on 12 September 1958, the first integrated circuit was born.

Within six months Robert Noyce of Fairchild Semiconductors (one of the traitorous eight) had a similar idea starting with somewhat different premises. He realized that the oxide of silicon is a tough material so it could serve as an external coat to protect the transistor from contamination. He later realized that metal strips evaporated on the top of the oxide layers could solve the problem of interconnections. Finally, he recognized the possibility of building resistors and capacitors out of silicon. The date when everything came together was January 1959.

All the developments, both at Texas Instruments and at Fairchild Semiconductors, were kept strictly secret. Kilby and Noyce did not know about each other's work. The application for a patent was first filed by Kilby on 6 February 1959. It was done in rather a hurry because some rumours (unfounded as it turned out) were afoot that RCA was soon to file a patent on the same thing. Noyce's patent application describing essentially the same thing but with much more detail on interconnections was dated 30 July 1959.

There was an inevitable patent fight, won by Kilby in the first round, but the appeal court and subsequently the Supreme Court found in favour of Noyce. The reason was the rather vague description of the interconnection scheme in Kilby's patent. Noyce's team claimed that interconnections cannot be made by *laying down* gold; the connecting material had to be *adherent to*.

But the mills of justice grind slowly. Patent fights take a long time. The final verdict was delivered as late as July 1969. Fortunately, the production of integrated circuits did not have to wait that long. In the summer of 1966, before the first court delivered its judgment, Texas Instruments and Fairchild Semiconductors reached an agreement to license each other for the production of integrated circuits and to co-license other companies.

There was no excitement when the first integrated circuit was presented at an exhibition alongside another 17,000 electronic products. There were only a few comments, all of them critical. For most people it seemed entirely unjustified to abandon the good old ways of producing resistors and capacitors. The frosty reception did not matter. Texas Instruments knew that they were on to something big, the biggest thing since the invention of the transistor. It was simply a question of convincing the biggest potential customer, the Pentagon, that the problem of the 'tyranny of numbers' had been solved. Texas Instruments started lobbying as soon as the first model was built. As may be expected there was initially a lack of interest, each of the services wishing to stick to their own pet projects. The first big breakthrough was a one million dollar contract from the Air Force for further development. The Air Force liked to claim subsequently how foresighted they had been to give immediate support to integrated circuits. In fact, they hesitated for a good six months and their main reason for support was that their molecular electronics approach did not seem to be leading anywhere.[6]

Robert Noyce at Fairchild Semiconductors was not in favour of military contracts. He thought they were too restrictive. So they relied on private capital instead in developing their integrated circuits. The actual outcome was, of course, the same. There was no civilian market at the time. Their integrated circuits were also sold to the military.

The capabilities of integrated circuits increased quickly with time. Gordon Moore, Noyce's colleague and another one of the traitors, was asked in 1964 how he expected the number of transistors in a single integrated circuit (they were called 'chips' by this time) to grow. He thought for a moment and recalled that in the previous three years the number of transistors had doubled every year, so he just said it would go on doubling every year. This off-the-cuff comment has been known in the industry ever since as Moore's Law. I remember quoting this to an economist friend of mine in the middle of the 1980s. 'How long has this been going on?', he asked sharply. 'About 23 years', I said. 'Are you suggesting', my friend's tone turned to derision, 'that you can now put 2^{23}, that is about 8 million, transistors on a chip?' For an economist sustained growth, even as little as 10 or 15 per cent per year, seems inconceivable. An increase by a factor of 2 for 23 subsequent years strained his credulity to the extreme. It took me quite some time to convince him that I was not having him on.

Let us pause here for a moment and look at actual figures. The doubling per year did indeed go on for some time but then it was reduced to 1.8 per year and even to 1.5. The basis on which the increase is counted has also changed. The first microprocessor saw the light of the day in 1971. It had 2,300 transistors. The 'transistor count' for each year from

[6] The properties of molecular layers are much better understood today, but there are still no working devices.

1971 to 2018 is shown in Fig. 10.2a. There is an increase from 2,300 to 2.3×10^{10} in 47 years. So we are now in the position to work out the increase per year. It leads to a factor of 1.41; i.e. the transistor count increased every year by 41 per cent. It is less than Moore's off-the-cuff prediction but still remarkable. If you put £1 in a bank in 1971 and received an interest rate of 41 per cent per year, you would have by now over 10 million pounds in your account (actually, less because HM Revenue and Customs would take their share). Another measure of progress is the 'minimum feature size'; i.e. how small is the smallest component? This is shown in Fig. 10.2b. The size declined from 10,000 to 7 nanometres, where 1 nanometre = one billionth of a metre. Compare this with the separation of atoms. The minimum feature size comes down to just 70 atoms. How long can this decrease go on? I don't know. Nobody knows. Perhaps a few more years, but then the decrease in size must stop. You can't possibly make transistors the size of an atom. Perhaps you can. Engineering ingenuity might come up with an entirely new solution.

Note that the transistor count went up by a factor of 10 million, whereas the minimum feature size shrunk by only a factor of 1,428. The answer is that this is the reduction in only one direction. If we consider the reduction in the other direction as well, then it comes to about 2 million. This is not the end of the story, because the chip size has also changed. It has increased in this time period by about a factor of 5, so counting this way we also arrive at 10 million transistors on a chip. Now back to the story of chips.

It was all very well to sell the integrated circuits to the military. That bore the expenses of early development. But surely, the great promise was to put integrated circuits into the home of every American. Considering that very few American homes were active in the guided missile business, something else had to be invented. There was one obvious candidate: hearing aids. For that purpose very small and very sophisticated amplifiers were needed. The snag was the rather small number of people with hearing difficulties. Were there any other consumer products which needed integrated circuits? There were none. So Patrick Haggerty, the president of Texas Instruments, decided to create a new consumer product, the pocket calculator.

Existing electrical calculators were expensive, big (just fitting on the top of a desk), heavy, and exasperatingly slow. The alternative was the slide rule, widely used by all those who wanted to have a quick answer. But whenever high accuracy was needed only those monsters were available. The pocket calculator was built by a team led by Jack Kilby. They managed to do all the arithmetical operations with no more than four chips. The first pocket calculators came out in 1971. They weighed a little over 1 kg and cost about $150. They sold like hot cakes; five million in 1972. Within a decade the price went below $10 and, at least in the US,

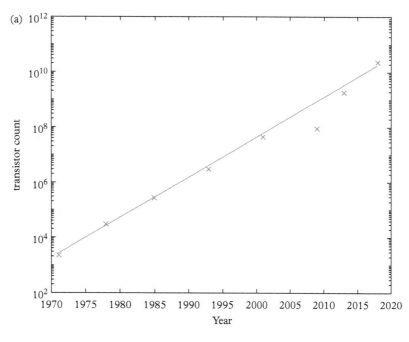

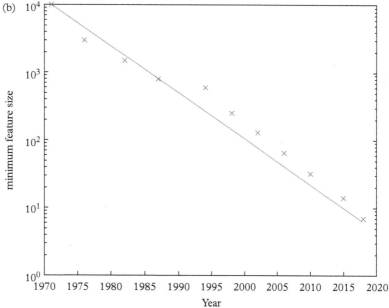

Fig. 10.2 (a) Variation of transistor count between 1971 and 2019. (b) Reduction in minimum feature size between 1971 and 2019.

the number of pocket calculators exceeded the number of people. Why did people buy them? They lived quite happily before without pocket calculators. The only possible reason I can think of is mass psychosis. The precondition is that a genuine need must exist, which might apply (say) to one tenth of a per cent of the population. However, the promotion campaign creates a strong urge among the rest of the population to imitate and to emulate. They will acquire the new gadget provided they know what it is for and, of course, the price must be within their budget.

Then came the digital wristwatch, vastly more accurate than its predecessor, for the bargain price of $275 (de-luxe versions going up to $2,000), followed by many other applications. Previously brainless things, like traffic lights, scales, lifts, and petrol pumps, suddenly became 'smart', meaning that they acquired some ability to respond to varying circumstances. But for each application a new circuit had to be designed. There was a definite danger of running out of circuit designers.

The solution to this problem was due to Robert Noyce, who meanwhile left Fairchild Semiconductors and founded Intel. His idea was to produce just one chip but make it programmable. So it became unnecessary to design a new integrated circuit whenever a new application emerged. It was sufficient just to reprogram the one and only design. In the beginning this chip was advertised as a whole 'computer on a chip', which of course it wasn't, but it could do all the processing. It is known nowadays as the microprocessor.

Naturally such chips, all being the same, could be produced at a low price. The first one to reach the market place was the 4004, which came out in 1971 and cost $200. It was followed by the 8008 and by the 8080, which long remained in production costing, at the end, as little as $3.

Next came the personal computer based on the microprocessor. People bought them up as they had earlier bought up pocket calculators. Most of the adults who bought them did not know what to do with them (apart from using them as typewriters), but their children did. No respectable household would nowadays deprive a child of playing games on a computer.

After personal computers came mobile phones, which brings us back to the subject of communications. They have been a success not only in terms of sales, but also because they do actually satisfy a need (see Chapter 14).

Digitalization

Analogue to Digital

An analogue signal, like voice, can take any value. Its amplitude is low when we whisper and high when we shout. The job of the telephone network is faithfully to reproduce all those voice signals, however long the line is. It's not an easy thing to do. The difficulty is with noise. It is bound to distort the signal. Take lightning, for example. It causes an electrical disturbance which clearly affects the amplitude of the voice signal. When the signal is demodulated and turned back into sound, a sharp, unpleasant click is heard.

Lightning does not occur very often. The main culprit is electrical noise caused by the random motion of the electrons. There is no way of persuading electrons to move in a nice uniform manner. In fact, a certain amount of noise is added from each point of the line, and quite a lot of noise from each amplifier. Thus, from the moment an analogue voice signal is generated in San Francisco to the moment it is demodulated in London, noise will keep on conspiring to corrupt the signal. Does this mean that a San Francisco–London telephone conversation is bound to be of low quality? Well, some lines are better than others but on the whole, yes, the conversation tends to be of low quality. Can it be improved? Yes, by using digital techniques any conversation between any two points on Earth may sound as good as a local call. How it is done is the subject of the present chapter.

How to Digitalize

Digitalization starts by sampling. If we sample an analogue waveform sufficiently often (see Fig. 11.1a), then, surely, the waveform can be represented as the sum of all the samples. The surprising thing, as shown by Harry Nyquist of Bell Laboratories in 1924, is that there is no need to take the samples very close to each other. The analogue signal can be reproduced by taking no more than two samples within the period of the fastest varying component of the signal. If, for example, the fastest varying component has a period of 1 ms, then taking samples at intervals of 0.5 ms will make it possible later to reproduce the signal from these samples. For the analogue signal of Fig. 11.1a this means that it is sufficient to measure the amplitudes at times t_1, t_2, t_3, and t_4, yielding the samples A_1, A_2, A_3, and A_4, as shown in Fig. 11.1b. Thus, instead of

Getting the Message: A History of Communications. Second Edition. Laszlo Solymar, Oxford University Press (2021).
© Laszlo Solymar. DOI: 10.1093/oso/9780198863007.003.0011

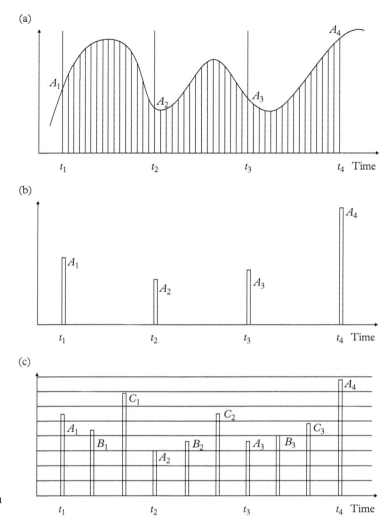

Fig. 11.1 (a) Sampling of a signal at close intervals. (b) Sampling at fewer intervals may be sufficient. (c) Samples from three different waveforms sent in succession: the basic principle of time division multiplexing.

sending the full waveform, we can get away with sending only a number of pulses.

Is there any advantage in sampling the waveform? Yes, the time between the pulses could be utilized for transmitting other pieces of information. In Fig. 11.1c pulses B_1, B_2, B_3 and C_1, C_2, C_3 have been inserted in the available space. They came by sampling two other waveforms. But this is nothing else than time division multiplexing, already mentioned in Chapter 4 in connection with Baudot's telegraph. So is this the modern way of sending audio signals? Not quite. There is one more step to take, as proposed by Alec Reeves in 1938 when he worked at the Laboratoire Matériel Telephonique in Paris. He was a very friendly, unpretentious, brilliant, extraordinary (and perhaps a little eccentric) man,[1] steeped in all aspects of telecommunications.

[1] He conducted experiments in telepathy and telekinesis under the name of Dr Soul.

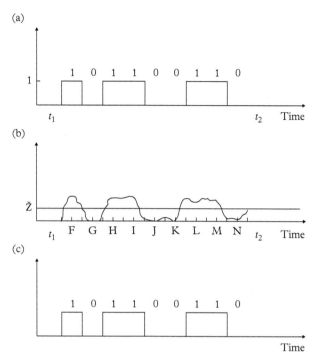

(a)

1   0   1   1   0   0   1   1   0

1

t_1

t_2 Time

(b)

Ž

t_1 F G H I J K L M N t_2 Time

(c)

1   0   1   1   0   0   1   1   0

Time

Fig. 11.2 (a) A signal amplitude expressed in binary form: the basic principle of pulse code modulation. (b) At the repeater the signal may appear severely distorted. (c) Reconstructed signal.

Reeves' idea was to express the amplitude in terms of binary numbers. In practice usually 256 levels are used, but just to illustrate the principles I shall use 8 levels here. Now A_1 is between 5 and 6.1 shall take the lower value, i.e. 5.[2] Similarly, I shall take 4 for B_1 and 6 for C_1. The next step is the crucial one. The amplitudes A_1, B_1, and C_1 are expressed as binary numbers. For those unfamiliar with binary numbers it may be worthwhile to recall that a number like 235 in the decimal system may be written as $2 \times 10 \times 10 + 3 \times 10 + 5 \times 1$. The decimal number is obtained by taking the coefficients of 10×10, of 10, and of 1 yielding 235. Similarly, in the binary system (which has only two digits, 0 and 1) a decimal number like 5 may be written as $1 \times 2 \times 2 + 0 \times 2 + 1 \times 1$. The coefficients 1, 0, and 1 yield the binary number 101. The decimal numbers 4 and 6 may be similarly expressed in binary form as 100 and 110, respectively. Coding now the binary 1 by a pulse and the binary 0 by the absence of a pulse, the digital signals to be sent for A_1, B_1, and C_1 within the interval t_1 to t_2 are shown in Fig. 11.2a.

At the receiving end the reverse operations are performed. The binary numbers are read and turned into the relevant amplitudes which, by using a technique known as filtering, can be converted back into the required analogue signal.

The modulation and demodulation of an analogue signal, as discussed in Chapter 5, was a much simpler affair. Why go to all the trouble

[2] Such quantization will surely introduce some error but if, as in practice, a much larger number of levels are taken the error committed is quite small.

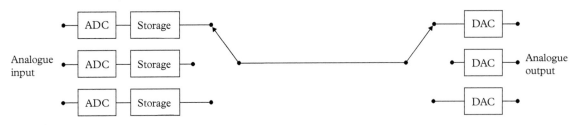

Fig. 11.3 Analogue signal converted to digital form and sent as a series of pulses down a line. At the other end it is reconverted into analogue form. ADC = analogue-to-digital converter. DAC = digital-to-analogue converter.

of converting to digital signals? The reason is that pulse code modulation (abbreviated as PCM) is much more resistant to noise.

One can't do anything about attenuation. Digital signals will decline along the line in the same manner as analogue signals do and will pick up noise in the same way. For example, the digital signal shown in Fig. 11.2a may be corrupted by noise by the time it reaches the other end of the line. It might be severely distorted as shown in Fig. 11.2b. Now comes the major difference between analogue and digital techniques. Analogue signals are amplified; digital signals are 'regenerated'. In order to regenerate a digital signal all we need to know is whether it represents a 1 or a 0, and it is a relatively easy job to determine that. We look at the signal at the middle of each interval (points F, G, H, etc. in Fig. 11.2b). If it is above half of the top value (as at F) we call it a 1, if it is below that (as at G) we call it a 0, and so on. We then send on to the next station the regenerated signal as shown in Fig. 11.2c. Since the set of binary signals are regenerated at each repeater the noise is not cumulative.[3] The net result is that a call from San Francisco to London does sound as if it came from next door. It is not only free of cracks but the voice of the caller can be clearly recognized.

For a practical realization we need converters. A schematic representation for three channels is shown in Fig. 11.3 where ADC and DAC stand for analogue-to-digital and digital-to-analogue converters. There are three analogue input lines. Each one of them is separately converted to digital signals and then they are stored until the right time when they can be released to the line. The electronic switches at the beginning and the end of the line need, of course, to be synchronized.

Digital Exchanges

As shown in Fig. 11.3 it is possible to send the signals in digital form on the line and convert them back into analogue form for the final destination—or for switching. There is no incompatibility there. Digital lines and analogue exchanges are perfectly capable of working with each other. One feels intuitively, however, that digital exchanges would be better suited to handling digital signals, and this is indeed the case. Digital exchanges are more reliable, faster, cheaper, take up less room, provide

[3] In fact, the noise can be kept down to an arbitrary low level (for discussion see 'Can We Get Rid of Noise Altogether').

better quality, and finally, make a new type of switching possible. It is called time-slot interchanging. It works in the manner shown in Fig. 11.4, again for three lines. A, B, and C are digital input lines and X, Y, and Z are digital output lines. The configuration is similar to that of Fig. 11.3 with one major difference. In the present case there is a time-slot interchanger in the line.

The two switches at the two ends of the line work in synchronism, If the order of the time slots is unaltered, then the signals from A, B, and C are routed to the output lines X, Y, and Z, respectively. There is now, however, the possibility of directing the signal from any input to any output by simply interchanging the time-slots. This does not mean that spatial switching has been entirely abandoned. In most modern digital exchanges a combination of time and space switching is used.

Bandwidth Requirements

How will digitalization affect bandwidth, the most valuable commodity for communications? Is more bandwidth necessary or can we get away with less bandwidth? One might expect the need for more bandwidth on the principle that 'there is no such thing as a free lunch'. If we gain on the noise front we shall surely lose on the bandwidth front, and this is indeed the case. The required bandwidth may be found by a few simple arguments.

The starting point is the highest sound frequency present in the signal. Taking it as 4 kHz, the sampling rate must be twice the highest frequency, i.e. 8,000 times in a second. The binary code will normally have 8 digits; i.e. each sampled amplitude is represented by 8 binary digits. Hence, the total number of digits sent in a second is 64,000. The fastest change is when a 1 and a 0 follow each other. That could be roughly represented by a sinusoidal wave, as shown in Fig. 11.5. The period of the

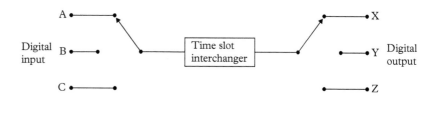

Fig. 11.4 Switching can be performed by interchanging the order of the pulses sent.

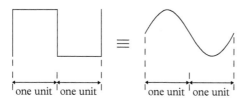

Fig. 11.5 When sending pulses the fastest variation occurs when a 'pulse' and a 'lack of pulse' follow each other. This sequence can be approximately reproduced by a sinusoidal wave of the same period.

sinusoidal wave is 2 binary units, and each binary unit is of a duration of 1/64,000 second. Hence, the corresponding frequency is 32 kHz. The conclusion is that PCM needs about 8 times higher bandwidth than single-sideband amplitude modulation. Is this a high price to pay for reduced noise? It all depends on the bandwidth available. If it is plentiful, go for PCM. If bandwidth is at premium (e.g. with microwave links), then it may make sense to stick to old-fashioned analogue signals. It is better to have a noisy telephone conversation than wait for ages for a connection to be established.

It is worth noting that although sound needs to be converted to digital signals, most other type of data (coming from faxes or computers) are already in digital form. Hence the converters may be eliminated but storage and switching are still necessary for time division multiplex. How many bits can be sent on a line in a second? It depends on the line. The main considerations are attenuation and dispersion.

Digital Signal Processing

There is always some signal processing involved when information is transmitted from point A to point B. Processing means interfering with the signal in a way that leads to a desirable result. Let me give an early example. It was found quite some time ago that, due to atmospheric conditions, the strength of a radio signal might vary considerably. A remedy was found by setting up automatic volume control. That meant that when the arriving signal was weak the amplification automatically increased and vice versa. It helped quite a lot.

The difficulty with analogue signals is that they exhibit such variety that each kind of processing necessitates a different circuit. The beauty of digital signals is that they are of only two varieties: 1 or 0.[4] An awful lot can be done just by storing them, rearranging them, or attaching some control signals to them in order to guide them to one place or to another. Let me start with an example that will drive home the power of digital signal processing.

For a local telephone conversation one line is sufficient: the voice can be carried in both directions. But if the other caller is far away, say at the other side of the Atlantic, then the line must have amplifiers in it which usually work in one direction only. Therefore, two lines are needed: one to carry my voice from here to America and the other one to carry my friend's voice from America to here. Now I want to ask a simple question: are two lines really necessary? Surely, in any telephone conversation only one person talks at a time, so the Oxford to New York line could be used by someone else while I am listening. It is a question of firstly recognizing whether I am talking and secondly switching the line to another pair of callers. It is worth doing if all this can be done in a time much shorter than the lengths of the silent periods.

[4] Digital signals are like an army in which there are two ranks only and they are capable and willing quickly to intermingle as soon as a command is given. Analogue signals are like a ragtag army in which everyone reacts differently to an attempt to regulate them.

Another example is bandwidth compression. The bandwidth of 32 kHz quoted earlier for a telephone line using PCM is not really necessary. The bandwidth needed could be considerably reduced due to the inherent redundancy in speech, i.e. due to the fact that sounds uttered in two consecutive time intervals are not independent of each other. They are correlated. This has been known for half a century at least, but the digital processing techniques and the necessary hardware have only recently become available at an economic price. The outcome of it is that a saving by a factor of 4 may now be possible; i.e. the pulse code modulated voice signal may not need more than 8 kHz, which is only twice that of an analogue signal.

For my third example I want to do away with the notion that a 1 is a 1 and a 0 is a 0. What I mean is that for digital communications we need two distinguishable signals for which a 1 and a 0 are often chosen so that at the detection stage they can be easily told apart. There is no reason, however, why these two signals should not take more complicated shapes containing, say, as many as 12 digits. The technique is known as direct sequence. Using this technique we may choose, for example, for signal number one 101101110000, and for signal number two 010010001111, as shown in Figs. 11.6a and 11.6b, respectively. Then, even if reception is very poor and one or two of the digits are misread, it is still easy to tell whether it belongs to one or to the other variety. Let us say the received sequence is 100 010 001 111 (Fig. 11.6c). Quite obviously it resembles signal number two more than signal number 1. But, to be sure, it is also possible to apply a simple test. How many of the received digits are at the same place as in signal number one, and how many as in signal number two?

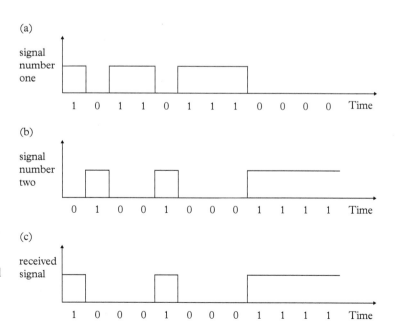

Fig. 11.6 (a) A possible representation of a logical '1'. (b) A possible representation of a logical '0'. (c) The received signal is not identical to either of the two above owing to severe distortion along the line.

Comparing the digits of the received signal with those of signal number one we can see that the first two digits are identical and all the rest (the next ten) are different. Doing the same comparison for signal number two we find that the first two digits are different but the rest are identical. So the decision is in favour of signal number two. Clearly, if instead of 12 digits we choose a sequence which is 100 digits long, then we can easily put up with 20 misread digits. Does this increased reliability come free? No, there is no such thing as a free lunch. If we send a sequence of 100 digits instead of 1, then the required bandwidth is 100 times as large. It is the job of the system designer to find the right compromise. If the requirements cannot be satisfied at a price the market will bear, then the system designer decides to wait, putting his (or her) faith in Moore's Law. With more elements on a semiconductor chip next year the same problem might be solved in a year's time more economically.

What else can be done? Speech recognition has been developed by now to a high standard.[5] It is possible to buy software that enables the running of computers by voice commands.

Digital Lines Enter Service (US, UK, and France)

The principles of pulse code modulation had been known since 1938 but the first digital line came into service only in 1962. Why did it take so long? From the idea to realization always takes some time but 24 years does seem a bit excessive in the communications industry. By the end of the Second World War it was generally accepted that digital communications was superior to the existing systems based on analogue signals. Nevertheless, there were no attempts to put it into practice. The reason was purely economic. The hardware needed was quite complicated, and to realize it with the existing electronic devices would have been very expensive. The advent of the transistor improved things but not sufficiently. Digital systems were still too expensive. To introduce them would have been similar to replacing an existing suburban railway network, much criticized by commuters, by a new one which is much more comfortable but nobody can afford.

In fact when digital lines first came, they came because they could do the job at the least expense. The motivation for building the first digital lines did not come from improving the quality of sound between San Francisco and London. It came as a solution to the problem of expanding services in urban areas.

By modern times all the telecommunications cables in the cities had been put under the streets, using a string of manholes for their maintenance and inserting new cables in the ducts when needed. Unfortunately, after a time the ducts inevitably filled up with cables. The only way to add new capacity was to dig up the city streets and lay new ducts. It was a prospect no telephone company (or, in the UK, the Post

[5] I remember the first attempts at speech recognition in the 1950s. The first problem to be solved was to recognize the numbers from 1 to 10. The man who worked on it proudly announced that everything works beautifully. When he utters the numbers from 1 to 10 the machine recognizes them without fail. I tried it, but it did not work for me. A few colleagues also tried. It worked for one of them. It turned out that the machine recognized only numbers spoken by people with a Polish accent. Today speech recognition is almost perfect. See, for example, the newsreaders' reports transcribed in real time on your television screen.

Office) contemplated with pleasure. It was too expensive. This is when digitalization came to the rescue. The proposed solution was to convert a single analogue line to a digital line capable of carrying 24 conversations.

The question that comes to mind is why resort to digital signals in order to carry multiple channels? Surely, a single cable can carry many analogue channels with the aid of frequency division multiplex, a technique known and used since the 1930s (see Chapter 7). Unfortunately, frequency division multiplex needed expensive, high-quality cables and expensive, high-quality amplifiers. It turned out, however, that PCM with time division multiplex was perfectly feasible on ordinary, low quality cables. This was due to the superiority of regeneration over amplification. The advantages were enormous. A line that up to then carried one conversation could suddenly be upgraded to carry 24. Admittedly, the terminals became more complicated, but for the lines the only additional expense was to replace the loading coils (there since the beginning of the century, see Chapter 5) with repeaters not larger than a matchbox, thanks to solid state devices.

Luckily, the progress of digitalization coincided with the progress in integrated circuits which kept the prices down. By 1984 there were more than 200 million circuit kilometres in existence in the US.

In the UK digitalization proceeded at a much slower pace. Switching in the post-war area was still dominated by antiquated Strowger type (electromechanical) exchanges which British industry kept on producing. In fact, the last one was supplied to BT in 1987 by the same factory which began producing them in 1912. The main concern of the British telephone companies was to maintain the same slice of the cake as was traditionally allotted to them by the Post Office. In any case there was not an awful lot of money going round because the Post Office was not allowed to raise money. The resources for investment depended on Parliament whose members knew little of the technical problems involved. It would have been the job of the civil service to push for modernization, but pushing for modernization is not the kind of thing the British civil service is good at and, of course, the Treasury is the last institution that would dish out money without serious arm-twisting.

As it happened there was some progress. Field trials started in 1966, and an experimental digital exchange was set up in 1968. Digital lines, first with 24 and later with 30 channels, were developed in the 1970s and installed in the junction networks, those connecting urban exchanges within the same city, in the early 1980s. Trunk lines were digitalized later in the 1980s. The complete digital system eventually developed is known generically as System X. It was developed by a consortium consisting of Plessey's, the General Electric Company (GEC), and Standard Telephones and Cables (STC). STC dropped out in 1982 owing to the fact that its parent company (International Telephone and Telegraph) had

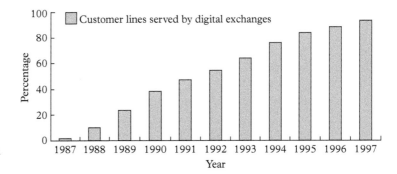

Fig. 11.7 Customer lines served by digital exchanges in the UK, 1987–97.

meanwhile developed a competing system. The rate of digitalization of the exchanges between 1987 and 1997 is shown in Fig. 11.7.

France is a rather special case. French telecommunications were the laughing stock of the world for a long time.[6] And then there was a sudden change. In the 1971–5 state plan, telecommunications was included for the first time as a national investment priority. In 1975 there were still only 7 million telephones in all of France and the waiting time for a new line was about 16 months. The great leap forward came during the presidency of Giscard d'Estaing. His advisers realized that the future of telecommunications was with electronics (although at the time practically all switching was done electromechanically), that electronics was closely related to computing, and that France needed a strong industrial base in both. Their advice was acted upon. By 1985 there were 22 million telephones in France and digitalization was in full swing. By 1986, 56 per cent of the network was digitalized, at a time when British Telecom had just about started on the job.

Why could the French do it? Because their government decided to do so. Is that enough? It occurs not infrequently that a government decides to build some new, glamorous infrastructure and it ends in failure. The French government and civil service are, however, better equipped than most to do this kind of thing. Thanks to the Grandes Écoles, which are their breeding grounds, they possess more technical expertise and are better aware of new technical developments than their counterparts in other countries. It is not entirely a coincidence that the president of the republic, Giscard d'Estaing, and his chief adviser on telecommunications, Jean-Pierre Souviron, were both graduates of the École Polytechnique, France's foremost school of engineering.

The Second Industrial Revolution Has Arrived

Let me summarize again the main steps, as I see them, in the development that brought about the modern age. To my mind it all started with William Thomson's calculations on the operation of the transatlantic telegraph cable. It was the first time that higher mathematics had been

[6] I remember a sketch on BBC television from the 1950s which showed a race in trying to deliver a message from Paris to Bordeaux. One man made the attempt by telephone, the other one by motor car. I cannot remember which one was the winner but the result was close.

used to solve an urgent engineering problem. It was the beginning of a trend that said that engineering intuition was fine but that it cannot always be relied on. Engineers had to accept that the solution of some design problems was provided by mathematical analyses of carefully set-up models, and that the reason for adopting certain design parameters rested on nothing more than the results of that mathematical analysis. If it was against common sense, well, common sense must have been wrong.

Next came Maxwell's equations. Out of the blue. And still they turned out to be correct. It was a great event in the history of science but, as it turned out, it was also a great event in the history of engineering. True, it took a generation from Maxwell postulating his equations to Marconi making the first use of electromagnetic waves, but eventually the engineers got there. Signals could be picked out of thin air.

And then came electronics. Scientists claimed that the electric charge was carried by a tiny particle called the electron. How tiny? How much smaller is the mass of an electron than (say) the mass of a marble of 1 cm diameter? It is smaller by a number that has 28 zeros in it. It is just out of this world. Engineers had no other choice but to get accustomed to working with things they could not see. Those tiny electrons, performing various functions in vacuum tubes, made possible worldwide communications. From the 1920s it became possible to establish contact between any two points on Earth. It was very expensive of course, but it could be done.

Next came the solid state revolution which still used those tiny electrons, but this time they were safely tucked into bits of solid material. Reliability went up. Prices plummeted. Then came computers which used those solid state devices, and finally there came the digital revolution that made it possible to send signals from one point to another point, wherever they were, with great flexibility and with arbitrarily high accuracy.

This is the end of the story. Neither muscular power nor brain power will be required in the future to any extent. We still need a few people to run the whole apparatus and a few more software analysts to find the causes of occasional hitches in the system, but that's about all. I shall return to that brilliant future (will it really be that brilliant?) in Chapter 22, the last chapter of this book.

Can We Get Rid of Noise Altogether?

A silly question, one might say. Whatever we do, noise will always corrupt our efforts. Oddly enough this is not what theory says. According to Claude Shannon it is perfectly feasible to obtain noise-free communications provided we are not too greedy; i.e. we are satisfied with a certain rate of transmission. This maximum capacity of a communication channel in bits per second is given by the formula

$$C = B \log_2 \left(1 + \frac{S}{N} \right)$$

where B is the bandwidth, S is the signal power, and N is the noise power. Taking for example a bandwidth of 32 kHz and a signal power to noise power ratio of one thousand, we find that it is possible to transmit information at a rate of 213 kbit per second without making a single error. How? This is where the problem starts. The proof of the above formula provided by Shannon in 1948 comes in the form of a so-called existence theorem. It proves that it can be done but does not tell us how it can be done.

Information theory, as this subject became known, is in the same category as some great theories of physics like relativity or quantum theory. It formulates a fundamental law. Shannon's formula was enormously influential in encouraging research into coding and noise-reducing schemes. If in principle we can get rid of noise altogether, then surely there must be means by which we can reduce it. And indeed a number of schemes sprang into existence. I shall mention here only one which to my mind is the simplest: the parity check.

Let us assume that we want to transmit the letter e, which in one code is represented by 1 010 001, a seven-bit code. The number of 1s in this particular code happens to be 3. Our aim is to make the number of 1s even. Hence we add one more 1 to the end. The letter e is sent as 10 100 011. The general rule is that if the code of a symbol contains an odd number of 1s, then a 1 is added; if it contains an even number of 1s, then a 0 is added as the last digit.

Why is this an error correction code? Because at the receiving end the parity (i.e. whether the number of 1s is even or odd) of each symbol is determined. If it is found to be odd then an error in the transmission must have been committed. The receiver signals back 'please, repeat the signal'. This method can, of course, be used only in a system in which the probability of an error is already low, because it would not notice two errors committed simultaneously.

There are, of course, many more sophisticated error-correcting codes. The final outcome is that errors can be kept very low indeed. If this wasn't so, many of our processes could not be automated at all. For example, automatic transfer of money between banks is an area where a low error rate is preferable. I believe the present standard allows on average 1 error in 10 billion signals. Another example would be releasing a nuclear attack. One hopes that the communications system which initiates that has a low error rate.

CHAPTER
TWELVE

Optical Communications: The Beginning

Introduction

Signalling with the aid of torches was, no doubt, a kind of optical communications and, of course, observing the positions of beams and shutters by telescopes may also be considered as belonging to the field of optics. A more modern way of using light for communications was invented by Alexander Graham Bell as an extension of his work on the telephone. He invented a device which worked on the principle that the human voice could modulate the intensity of light. The modulated light then propagated through air to reach the line-of-sight receiver, where the varying light intensity gave rise to a varying current, which in turn could be converted back to sound. The device was called a 'photophone'. It worked well in good weather. Contemporary comments were not all admiring. A *New York Times* editorial in 1880 expressed the opinion:

Does Professor Bell intend to connect Cambridge and Boston with a line of sunbeams hung on telegraph posts, and if so what diameter are the sunbeams to be . . . what will become of the sunbeams after the sun goes down? . . . The public has a great deal of confidence in Scientific Persons, but until it actually sees a man going through the streets with a coil of No. 12 sunbeams on his shoulder and suspending it from pole to pole, there will be a general feeling that there is something about Bell's photophone which places a tremendous strain on human credulity.

The principles were sound but the photophone was never a practical proposition. Similar ideas appeared occasionally over the next 70 years but none of them was of any import until the concept was taken up towards the end of the 1950s by Alec Reeves of Standard Telecommunications Laboratories, whom we have already met as the inventor of pulse code modulation. He was not only steeped in all the techniques of telecommunications; he was a visionary as well with an instinctive feel for the right approach leading to the right solution. He was convinced that the eventual winner in the long-distance communications race would be optics.

Why Optics?

Of all the electromagnetic spectrum the part most studied in the course of human history is light. The main reason has been the early availability of powerful light sources ('Let there be light', Day 1) supplemented

Getting the Message: A History of Communications. Second Edition. Laszlo Solymar, Oxford University Press (2021).
© Laszlo Solymar. DOI: 10.1093/oso/9780198863007.003.0012

five days later by a marvellously effective broadband light detector: the eye. Ever since, the subject has been a playground of philosophers and physicists, of theoreticians, and of men of practical bent. The amount of knowledge accumulated by the 1950s was formidable. By that time there were light detectors superior to the eye, man-made light sources which were powerful enough to compete with the Sun, and means to modulate the intensity of light.

So one could simply use the principles of the photophone: take a light source, modulate it, send it some distance, and demodulate it. The new thing would be to shield the beam, e.g. put it in a pipe so that nobody can interfere with it, and make the modulation digital, e.g. turn the light source on and off by using a fast shutter. Such a simple arrangement would be an efficient way to communicate, provided the light source could be modulated fast enough. If light pulses of 1 nanosecond (1 billionth of a second) duration can be produced, that would make possible 15,000 high-quality telephone conversations (remember, one channel needs 64,000 bits/s) by using some guiding structure, e.g. a shiny metal pipe in which light can bounce off the walls. Research on roughly these lines started in 1958 under the direction of Alec Reeves.

Meanwhile some physicists started to think about optical sources in a novel way.

The Laser Appears

Advances usually come in small steps. A swimmer might be celebrated if he improves the world record from (say) 3 minutes 25.2 seconds to 3 minutes 24.3 seconds although the change in the time clocked is a mere 0.4 per cent. The largest ship, the tallest building, the fastest aeroplane when they first appear might beat the previous record by a factor of 2 but not much more. Advances in communications can actually be much steeper. When Marconi experimented with short waves in the 1910s the jump in frequency was by a factor above 10. About the same factor of 10 was achieved by the war-time efforts when microwave radar was introduced. So a new jump towards higher frequencies might well have been expected in the late 1950s. What was the state of the art then? The highest frequency available was about 150 GHz. At the time the wildest expectation might have been for a further increase by a factor of 7 or 8. The research was fuelled by the military requirements of electronic measures and countermeasures.[1]

And then out of the blue came a theoretical proposal to reach frequencies a couple of thousand times higher. Two American physicists, Arthur Schawlow and Charles Townes, claimed that it would be possible in principle to produce electromagnetic waves with frequencies of about 500 THz.[2] The corresponding wavelength is 0.6 μm, which is well known to cause the sensation of red colour in the brain. In other words the new

[1] An indication of the interest of the US Air Force at the time is that they were willing to support research for shorter millimetre waves at the French electronics company CSF while de Gaulle was delivering his anti-American tirades.
[2] The new unit appearing here is T for 'tera' meaning 10^{12} Hz = 1000,000,000,000 Hz = 1 million MHz = 1,000 GHz.

proposal envisaged producing visible light. OK, you might say, what's so great about that? We have had light of all colours ever since the Creation. The difference is that the prediction was about producing coherent light.

Perhaps the simplest (although not quite accurate) way to define coherent waves is to say that the radiation is of a single frequency. The waves coming from the Sun contain a wide range of frequencies, as witnessed by rainbows. On the other hand the electromagnetic waves which bring us radio and television are close to a single frequency. For incoherent waves the 15,000 channels mentioned earlier would have been a marvellous achievement. For coherent waves, for which our old calculations apply, it's just peanuts. Taking a 15 per cent bandwidth around 500 THz the number of telephone channels comes to about 18 billion with single-sideband modulation, and still above a billion by pulse code modulation. This means that if we managed to concentrate all the European population into a phone box in London and all the Americans (North and South) in a similar phone box in New York, then they could all talk to each other on a single line and there would still be some spare capacity.

Oscillators producing electromagnetic waves always worked on classical principles. The proposal by Schawlow and Townes was based on quantum physics which is an exceedingly obscure subject. Fortunately, for oscillators, the quantum theoretical aspects can be very simply summarized by the maxim: 'what goes up, must come down'. The 'going up' means that energy is given to the system. The 'coming down' has to be a little more specific. The energy gained has to be given out in the form of radiating electromagnetic waves.

In 1960, less than two years after the appearance of the paper by Schawlow and Townes, the experimental proof was provided by Theodore Maiman in California. He managed to obtain coherent red light with the aid of a piece of ruby crystal illuminated by a flashlamp. The new device came to be called (not immediately, but a little later) a laser, which is an acronym for light amplification by stimulated emission of radiation. In fact, the laser is more of an oscillator than an amplifier, but the corresponding acronym 'loser' just did not sound right. (I should perhaps mention here that according to some people the acronym laser has a much earlier origin, as suggested in Fig. 12.1.)

It needs to be emphasized that the laser did not appear in response to industrial demand. At the time resources for scientific research were plentiful and no justifications were needed to prove the immediate or even eventual usefulness of the research. Perhaps the greatest spur was the human desire for exploration. Mount Everest had to be climbed because 'it was there'. The North Pole and the South Pole had to be discovered and space had to be explored in order to satisfy human curiosity. Well, the exploration of the electromagnetic spectrum was

Fig. 12.1 Light and smoke emission regulator, or laser.

just part of this general zealousness; it was done for the greater glory of science.

Optical Fibres for Communications?

For Reeves the advent of the laser came as an unexpected boon. His men were already running along the right track when a bandwagon suddenly appeared. They jumped upon it without losing any time while the rest of the world was not even aware that there were any bandwagons on the move. Reeves and his co-workers looked at various guiding structures: the hollow metal tube already mentioned; an array of lenses (called confocal waveguides because the light in them was focused and defocused as it propagated from lens to lens; see Fig. 12.2); and the dielectric waveguide which had already existed in the form of bundles of glass fibre and which had been used for medical purposes. A dielectric is nothing more than an insulator, a material that does not conduct electricity. The mechanism by which they could guide electromagnetic waves had been known for a century at least. For a qualitative explanation see 'Dielectric Waveguides (see p. 222)'.

Armed with the knowledge that coherent light had become available, did people in the field realize that optical communications was around the corner? By that time it was an obvious assumption. In all the previous history of the subject whenever generators of higher and higher frequencies were produced, they were invariably used for communications. The question was not whether, but when. I remember discussions at the time at Standard Telecommunications Laboratories.

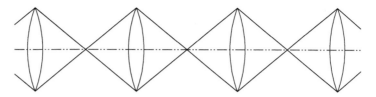

Fig. 12.2 A confocal waveguide: light propagates by focusing and defocusing.

Most of us believed that the millimetre waveguide would come in the 1970s and optical communications in the 1980s. What form did we think optical communications would take? The answer depended on the background of those making the prediction. Those without a background in electromagnetic theory thought that a metal pipe with silver deposited on its inner surface might do. Those who came to the problem from the optical end were in favour of the lens array of Fig. 12.2, and finally those with a microwave background thought of dielectric waveguides as the likely winners.

Toni Karbowiak, who was by 1964 in charge of the transmission medium studies at Standard Telecommunications Laboratories, concluded in a paper read to the Institution of Electrical Engineers that of all the guides known to-date the fibre guide appeared to hold most promise, if due to advances in material technology, it became possible to manufacture cladded fibres with effective loss tangent about two orders of magnitude better than at that present time.

Cladding was an important part of the setup. The fibre diameter needed for single-mode operation was comparable with the wavelength, that is, on the order of one-thousandth of a millimetre, much thinner than a strand of human hair.[3] It was to be kept in place by the cladding, which also had to have low losses since light penetrated it to a certain extent.

The main problem was losses. The glass fibres available at the time gobbled up the power. The best rate for attenuation was 1 decibel per metre. That meant that the power declined to 10 per cent of its value after travelling one metre in the material. This is quite a sharp decline. After travelling a mere 120 m the light emerging would be only one billionth of that entering. Quite obviously some considerable improvement was necessary before fibre guides could be used for communications.

Karbowiak left STL in December 1964 to take up a professorship in the sunnier climate of Sydney. The remnants of the group, Charles Kao and George Hockham, continued the good work. Looking at the state of the art they concluded that all the likely obstacles (scattering of light by impurities, attenuation, effect of bends) seemed to be surmountable. In particular, there was an encouraging report published in the previous year by F. N. Steele and R. W. Douglas claiming that an attenuation of better than 20 db/km was achievable if the iron impurities in glass could be reduced to 1 part per million. The situation looked moderately

[3] Single-mode operation means that the field profile (light is an electromagnetic wave, so it has an electric field!) does not change as the wave propagates.

hopeful. The likely reason for the presence of impurities was that nobody ever tried to get rid of them. There had been no motivation to do so since the existing glasses satisfied all demand—but then nobody had wanted to use fibres for communications before.

Millimetre waveguides had attenuation of a few decibels per kilometre at most. For optical fibres to become competitive on trunk lines the same kind of attenuation would have been needed and that was entirely out of the question at the time. There was, however, another kind of application for which optical fibres were suitable but millimetre waveguides were not. Let us remember that the initial motive for digitalization was the fact that the ducts in the junction lines were full. Capacity between urban exchanges was increased by converting lines carrying one analogue channel to carriers of 24 digital channels, thanks to the technique of pulse code modulation and time division multiplex. Optical fibres could, of course, do the same thing even better. They were flexible, they could carry digital modulation, and they would fit nicely into ducts. If the attenuation was low enough (and about 20 db/km was deemed to be sufficient), then the next repeater could be put into the next manhole and the whole system would work beautifully.

Low-Loss Fibres

If the target is too far away, the incentive to do research on it tends to be fairly weak. If there is an intermediate aim as well, then that improves the chances. So the hope that optical fibres might be used in junction lines made laboratories keener to initiate some research. However, everything started in a low key way because the technological problems to be solved were very difficult indeed. A low-loss fibre with a low-loss cladding adhering smoothly to the core was needed,[4] and the cladding had to have an index of refraction somewhat lower than that of the core. The difficulties of producing a low-loss fibre were formidable enough without having the additional problem of the right kind of cladding.

Glass fibres were usually produced by melting glass in a crucible and pulling it through a hole at the bottom. The speed of the pulling determined the diameter of the fibre. Of course, some care had to be exercised so as not to break the fibre, but the technology was there. How to change the index of refraction was also known. It could be changed by adding certain materials to the melt.[5]

The companies involved in producing fibres were mainly telecommunication companies (Standard Telecommunications Laboratories in England, Nippon Telegraph and Telephone in Japan, Bell Laboratories in the US, Compagnie sans Fils in France), the British Post Office, a few universities (notably the University of Southampton), and a few glass companies. The one that succeeded was Corning Glass Works in the

[4] Any surface roughness would add to the losses by scattering the light.
[5] These materials are usually referred to as dopants and the operation is known as doping.

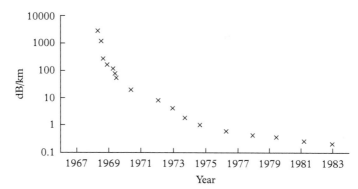

Fig. 12.3 The best values of optical fibre attenuation in the period 1967 to 1983.

state of New York. The manufacturing process that led to success was the use of titanium doping.

The head of the group, Bob Maurer, attended a conference in London later in the year where he announced the production of the lowest loss glass ever produced by man. A glass company is in the business of selling glass. In order to sell their latest product they had to convince potential users that their product was good. Bob Maurer decided to have independent confirmation of his attenuation results by allowing the British Post Office and Standard Telecommunications Laboratories to make measurements of his fibre. He wanted the confirmation but without giving away the secret of how the cladded fibre was made. He made sure that the fibre and himself never departed from each other. He even refused to have lunch in the canteen at the Post Office's Dollis Hill Laboratories in order to guard his fibre. He was happy to eat sandwiches after the lunch party returned.

After the first triumph Don Keck and Peter Schultz continued the good work at Corning. Replacing titanium with germanium they obtained an attenuation of 4 db/km by June 1972. Now optical fibre had a low enough attenuation, not only for junction lines but for trunk lines as well. The effort to produce clad fibres with even lower attenuation gathered further momentum. The attenuation achieved as a function of time is shown in Fig. 12.3. It represents, no doubt, one of the greater triumphs of materials engineering. In order to drive home how small the achieved attenuation is, let me note that if the sea were that transparent we could easily see the bottom at its deepest point.

Light Sources

It is not enough, of course, to have low-loss fibres. For optical communications, sources are needed too. With the advent of the laser, light sources did indeed become available but they were rather big, clumsy, and unreliable. They were not the kind of thing the Post Office would have wanted to put in a manhole. What was wanted was a small, rugged, and reliable laser. Was there a way to produce one? There was. The semiconductor laser was invented in 1962 by as many as four American groups quite independently and practically at the same time: no more than a month or so between them.

It was nice to have a semiconductor laser but it had three major deficiencies at the time. It had to be operated at liquid nitrogen temperatures ($-196°C$), it could only work in the pulsed regime (i.e. on for a short time, off for a longer time), and it had a very short and unpredictable life. It was in no condition to be the source for optical communications.

The next advance came in 1967. By using different materials and making further improvements to the junctions a source that can be used for fibre communications was born. The idea was to have different materials on the two sides of a junction (aluminium gallium arsenide on one side, gallium arsenide on the other side), making a so-called heterojunction. The advantage was that the electrons and holes could preferentially congregate in one of the materials, thereby helping the process of recombination. It did indeed help, but not enough. The next idea was a double heterojunction, meaning one narrow piece of gallium arsenide sandwiched between two pieces of aluminium gallium arsenide. It concentrated the electrons and holes even more in the gallium arsenide in the middle, but it was still not enough. Then came the idea of masking off part of the top of the laser with an insulator so that the current was concentrated in the middle of the device. The semiconductor laser designed this way could produce continuous radiation at room temperature. The laboratory that first created the semiconductor laser was the Ioffe Institute in Leningrad. The Americans were not much behind. Bell Lab announced their first stripe geometry laser a month later. The remaining problem was reliability. How to increase the lifetime of a laser? In mid-1977 Bell Laboratories announced lasers with an expected lifetime of a million hours,[6] that is, over 100 years. These were lasers good enough to be put in cables under the ocean.

Demonstrations and Trials

The first public demonstration took place on 19 May 1971 when Murray Ramsay showed the Queen the transmission of television signals via an optical fibre (see Fig. 12.4). The recipient of the first fibre system that could actually do something useful was a Dorset police station where the previous apparatus was struck by lightning, and which was looking for a quick replacement. The engineers of STL obliged. They connected up the video display units with the aid of optical fibres.

[6] How does one know? One can't sit there watching the laser for a million hours. The technique is to test the laser under 'accelerated' conditions (i.e. at higher temperature or current) and predict from the lifetime found the likely lifetime under normal conditions.

Fig. 12.4 Murray Ramsay showing television transmission by fibre to the Queen.

A substantial trial of the capabilities of optical fibres was conducted by Bell engineers in Atlanta in 1976, followed a year later by the trial of a 2.6-km fibre cable under actual field conditions in the murky manholes of a Chicago district. It was an unqualified success. The fibre cable could carry several hundred times more channels than a copper cable.

In Britain trials by the Post Office in Martlesham and by STL between Stevenage and Hitchin took place in 1977. There were occasional hitches but there was no longer any doubt that the introduction of fibre lines in service was only a question of time.

The start-up was not particularly fast. A newcomer like optical fibre had to find its niche in the vast telecommunications network that already existed. The first commercial line was built between Boston and Washington, carrying traffic at 270 million bits per second per fibre. When it opened in 1984 it was already obsolete. By this time the Bell System had lost its monopoly. Their rivals had the advantage of starting from scratch. To lay optical fibre lines was one of their first priorities.

Major changes took place in Britain as well. First, the telecommunications interests were hived off from the Post Office, and then British Telecom became a separate company. They were so proud of their achievements in fibre optics that the length of optical fibre lines installed got into their annual reports to the shareholders. The spectacular growth between 1986 and 1997 is shown in Fig. 12.5.

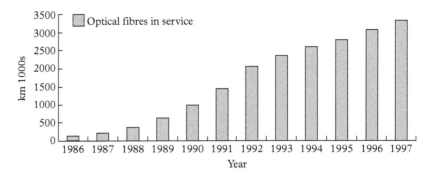

Fig. 12.5 Optical fibres in service in the UK in the period 1986 to 1997.

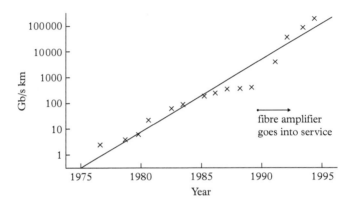

Fig. 12.6 The best values of capacitance–distance product in the period 1975 to 1995.

The Fibre Amplifier

Thanks to the technological effort invested, the attenuation of optical fibres was reduced to a very small value indeed. That meant that repeater spacing could be as much as several hundred kilometres, but even then lots of them would still be needed to span the Atlantic or the Pacific Ocean. The disadvantage of electronic repeaters (i.e. conversion from optical to electronic form, reconstruction and amplification of the signal, and finally modulation of the optical carrier) was that they could only work at a fixed transmission rate, whatever the demand. The next stage of development took care of this problem by making the repeater superfluous.

The solution was to employ amplifiers, but not just any kind of amplifiers. The beauty of the solution was that the amplifier was also part of the line, just a bit of fibre doped by a material called erbium and illuminated by another laser. It was first produced in 1987 by a team at Southampton University led by David Payne. It came just at the right time to give a boost to fibre performance. This may be seen in Fig. 12.6, where the achievable capacity–distance product is plotted for the 20 years period from 1975 to 1995.[7] It has been doubling practically every

[7]The capacity–distance product is a measure of how far a signal carrying so many bits per second can travel without regeneration. The highest figure shown is about 2×10^3, which may be interpreted, for example, as the ability to deliver 40 Gb/s (equivalent to about 60,000 telephone channels) to a distance of 5,000 km.

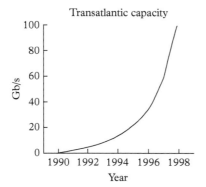

Transatlantic capacity

Fig. 12.7 Transatlantic capacity in the period 1990 to 1998. (Note that 1990 is not zero.)

year, except for a five-year period when it was fairly steady, only to rise again suddenly with the advent of the fibre amplifier.

Fibres under the Seas

The first telephone cable across the Atlantic was laid as late as 1956. It had only 35 channels, not a lot if we think of the potential traffic between Europe and North America. By 1988 another 8 conventional lines were laid but their total capacity hardly exceeded 10,000 channels. Then, at the end of 1988 the first transatlantic fibre cable went into operation, suddenly quadrupling the existing submarine capacity. The growth of capacity from 1990 to 1998 is shown in Fig. 12.7. A capacity of 100 Gb/s corresponds to about 1.5 million telephone channels.

There were about 50,000 telephone channels available at the time but it hardly shows on this scale. Having said a lot in practically every chapter of this book about the capacity of cables, I have to admit here that the concept has become a little outdated. Once upon a time the capacity was determined once the cable disappeared below the surface of the water. With the invention of fibre amplifiers all that has changed. The capacity can now be varied by simply changing (or adding to) the terminal equipment, without touching the cable at all. One of the new techniques is wavelength division multiplex, which means simply that another carrier frequency, to be modulated by digital signals, is added.[8]

Another technique which makes it uncertain what is the total number of telephone conversations that could be carried on a line is bandwidth compression, already mentioned in Chapter 11 on digitalization. The necessary bandwidth might be reduced by a factor as large as 4 or possibly 8 with hardly any effect on quality. It is made possible by the redundancy in human speech: whatever we say in the present millisecond is not independent of what we said in the previous millisecond. It is a technique worth using when bandwidth is at a premium.

[8] It is exactly the same thing as frequency division multiplex discussed in Chapter 7, but for some reason it got a new name.

Optical Fibres versus Millimetre Waveguides

In Chapter 9 I wrote briefly about the efforts to use millimetre wave-guides on trunk lines capable of transmitting hundreds of thousands of telephone channels. By 1970 the optical fibre was a strong contender for those trunk lines. Indeed, by that time, Richard Dyott of the Post Office Research Laboratory would taunt his comrades in the millimetre wave-guide camp, saying, 'I'm quite happy for you to lay the waveguides, and we will come along later and fill them with optical fibres.'

Were optical fibres the predestined winners? Eventually, yes, but not necessarily in the short run. Had the Picturephone, launched in 1970 in the US, been a success, the demand for bandwidth would have risen sharply and the millimetre waveguide would have been called into service. Once in service, all technologies put up a tremendous fight before accepting their own demise. There would have been scores of managers egging on research engineers all over the world to improve the existing product and fend off the intruder—optical fibres. For example, the problem caused by movement of the soil (thus making the waveguides deviate from perfect straightness) would surely have been solved by inserting better mode filters in sufficient number. The demand for more capacity for junction lines could also have been met without optical fibres by continuing the conversion to digital lines. Optical fibres would have come eventually, but perhaps the first optical transatlantic cable would not have been laid until the turn of the century.

Dielectric Waveguides

How to guide the wave

The phenomenon of light refraction has probably been known since the start of human civilization. Put a stick in water and it appears as if it is broken. It is a consequence of the fact that when light moves from a medium of lower refractive index (like air) into a medium of higher refractive index (like water) the rays will change direction at the boundary as shown in Fig12.8a. When light is incident from a higher refractive index material it is refracted towards the boundary. As the angle of incidence increases, there is an angle at which the refracted ray is directed just along the boundary. This is when the refraction turns into reflection, a phenomenon known as total internal reflection (see Fig. 12.8b). It follows then that if light is incident at the surface of a glass rod (which has an index of refraction higher than that of the surrounding air) at a shallow angle, as shown in Fig. 12.8c, it will be trapped inside by subsequent reflections. Such guidance of light waves was probably known to the ancient Egyptians who used glass tubes for ornamental purposes. Note the similarity between Figs.12.8c and Fig.9.6. In both

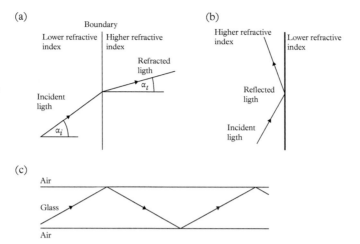

Fig. 12.8 (a) A ray of light refracts when moving from a medium of lower refractive index into a medium of higher refractive index. (b) If the angle of incidence is small relative to the boundary, then there is no transmitted ray in the medium of higher refractive index and the ray is completely reflected. (c) Light may travel in a glass rod by subsequent reflections at the boundaries.

cases the electromagnetic wave propagates by multiple reflections. The difference is that in the former case the wave propagates in a dielectric, whereas in the latter case the wave propagates in air and the reflections are caused by metal boundaries.

If the aim is to guide the light along the dielectric waveguide, then it is strictly necessary for the index of refraction inside to exceed that outside. Since the guide cannot levitate in air and is in need of some support, this condition will not apply at the points of support where the electromagnetic energy may leak out. The way to overcome the problem is to clad the dielectric waveguide in a suitable material. It turns out to be advantageous to choose a cladding material whose index of refraction is just slightly smaller than that inside. Such an arrangement would allow single-mode operation with a reasonably sized, not too small, inner core.

History

The first demonstration of light trapping was done more or less simultaneously by Daniel Colladon in Geneva and Jacques Babinet in Paris. In Colladon's experiment, conducted in 1841, light was made to move along water jets and break free when the jets themselves broke, providing some spectacular pictures to the great enjoyment of the public.

One of the ways of producing a dielectric waveguide is to draw molten glass fast enough so that it turns into a thin glass fibre. This was done by Charles Vernon Boys towards the end of the nineteenth century at the Royal College of Science in South Kensington. His method was original, to say the least. He stuck an arrow to one end of a glass rod, which he heated until it was soft enough, and then shot the arrow to a distance of

90 ft. He was rewarded by a fibre 90 ft long and having a diameter of one-ten-thousandth of an inch.

A more modern embodiment of a dielectric waveguide came in the 1950s with the endoscope, which consists of a bundle of glass fibres no longer produced with the aid of bows and arrows.

CHAPTER THIRTEEN

Deregulation and Privatization

The Break-up of the Bell System in the US

The end of the Second World War saw AT&T stronger than ever. They had the local monopoly, the long-distance monopoly, and the manufacturing monopoly. Almost all the equipment used by the AT&T network was manufactured by Western Electric, their own company. The US Justice Department thought that was too much of a good thing and filed a suit in 1949 seeking to divest Western Electric from AT&T. The case dragged on until 1956 when an out-of-court settlement known as the Final Judgment was reached. AT&T was allowed to retain their manufacturing arm in return for a few minor concessions, namely that (i) Bell companies would license their patents under reasonable and non-discriminating terms to all applicants,[1] and (ii) they would not engage in any unregulated business. This meant that anything related to telecommunications was all right but anything outside, e.g. computers, was forbidden.[2]

The next assault of the Justice Department, which always kept a wary eye upon AT&T, started in 1974 under the presidency of Gerald Ford. The armour of AT&T appeared as strong as ever but those in the know could see some chinks due to the assault of advanced technology.

It all started with microwave links. If a company had a lot of business at a site some distance away from headquarters, they would set up a private microwave link between the two places. As mentioned in Chapter 9 (microwaves), such a venture needed only a moderate amount of investment: to build a few towers and buy the equipment from independent manufacturers. Anybody could set up a private line so AT&T could not object. However, according to the existing legislation, a private line had a wider meaning. It was not restricted to connecting two specific sites of the same subscriber or connecting two independent subscribers. It was also permissible to connect two public exchanges. If someone, say from St Louis, needed frequent connections with various subscribers in the Chicago area, he could rent a private line from the St Louis exchange to the Chicago exchange. His calls to any number in Chicago would then be regarded as local calls.

The facilities offered by such a private line (called a foreign exchange) may be represented by Fig. 13.1a, which shows that subscriber A from St Louis could call any subscriber in Chicago. It was, of course, possible to do the reverse. Subscriber B from Chicago could also rent a private line

[1] An earlier example of this was the licensing of transistor manufacture at a moderate rate.
[2] To abandon any activity in the computer field appeared at the time no more than a minor concession. However, that concession alone would have sooner or later undermined AT&T's dominance in the telecommunications business.

Getting the Message: A History of Communications. Second Edition. Laszlo Solymar, Oxford University Press (2021).
© Laszlo Solymar. DOI: 10.1093/oso/9780198863007.003.0013

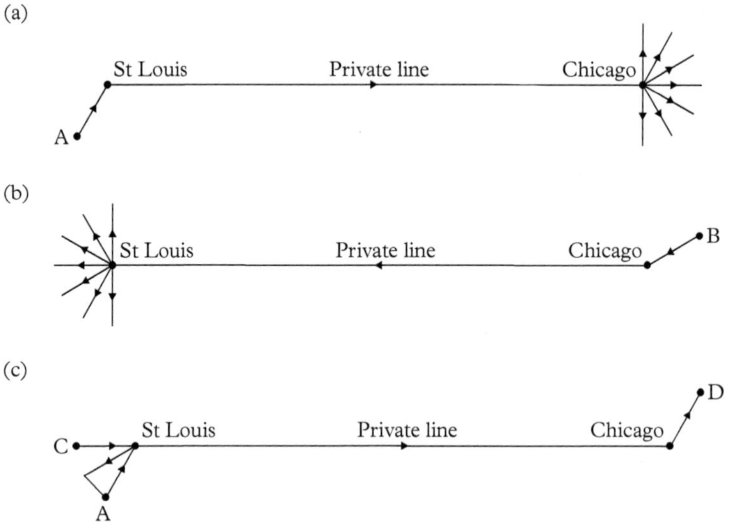

(a)

(b)

(c)

Fig. 13.1 'Foreign exchange' facility, (a) By hiring a private line a subscriber from St Louis may call any subscriber in Chicago. (b) By hiring a private line a subscriber from Chicago may call any subscriber in St Louis. (c) Using both facilities any subscriber from St Louis may call any subscriber in Chicago, and vice versa.

which enabled him to call any St Louis subscriber at the local rate. This is shown in Fig. 13.1b. Now what happens if a company decides to rent two foreign exchanges, one open at the Chicago end and the other one open at the St Louis end? That would enable the company to offer a St Louis–Chicago telephone service; i.e. any subscriber from St Louis could call any subscriber in Chicago and vice versa.

An example is shown in Fig. 13.1c. Subscriber C in St Louis may call subscriber D in Chicago by first calling A on the local network and then, relying on A's access to any Chicago number, he can call D via A. This can, of course, be made automatic at the expense of dialling a few more numbers. Thus any company could challenge the long-distance monopoly of AT&T by setting up at least two private lines between exchanges in different cities.

Figure 13.1 shows that competition in providing long-distance connections was technically possible, but was it profitable to do so? Could any company undercut the rates of AT&T? The answer was yes, they could, because of AT&T's pricing policies. There was nothing odd about those policies. They were in line with AT&T's desire to provide a universal service. It meant that the charges for long-distance calls were the same, whether they were between rural areas (with just a few conversations going on) or between busy urban centres. The charges depended only on the distance. Thus, clearly, urban subscribers subsidized rural subscribers. Another cross-subsidy was mainly the result of state regulation. The aim of the regulators was to keep local calls as cheap as possible. After all, there were many more people making local calls than long-distance calls, and each subscriber had a vote. Thus, whichever party was in power wanted to stay in power by continuing the subsidy.

The outcome was that long-distance callers between urban centres were charged a lot more than the economic price. Thus, it made good sense to rent private lines from AT&T or go even further and build a microwave link between the two cities. Here was an opportunity to 'skim the cream', i.e. to compete for the favours of business users. This was indeed the strategy of a small company called MCI (Microwave Communication Inc.). They set up private lines with foreign exchange facilities and asked permission from the Federal Communications Commission to operate them. AT&T opposed the applications tooth and nail. They might have won if judge X rather than judge Y had been in the chair on some occasion or another but their victory would have only been temporary. The writing was already on the wall. AT&T was weighed in the balances and was found wanting in the field of competition. With the advent of microwave links, the natural monopoly theory could no longer hold.

The first judgment against AT&T was made in 1968 in the so-called Carterfone case. Thomas Carter was a Texas businessman providing private lines for oil companies. In order to help his customers, he concocted a device to enable them to connect their private phones into the public telephone service. AT&T was not amused. They took Carter to court but they lost. The court decided that interconnecting equipment was permissible providing it did not adversely affect the system.

The AT&T versus MCI case dragged on and on, nearly bankrupting MCI in the process. In 1975 MCI was banned from using foreign exchange facilities but the judgment was reversed two years later. The saga ended with MCI's complete victory. Long-distance telephony became deregulated. Any company could enter the arena. AT&T was required to divest itself of all the local companies which would remain regulated monopolies. In return AT&T was free to engage in any unregulated activity, including computers. The agreement was announced to an incredulous public in January 1982. Since it was a modification of the agreement of 1956 it became known as the Modification of Final Judgment.[3]

The divestiture took place on the first day of 1984. Seven regional Bell operating companies, dubbed Baby Bells by the popular press, were set up. In contrast to the expectations of many, the world did not come to an end. Everything went on as before. The telephones kept on buzzing but AT&T, deprived of its monopoly power, became just another company striving to make profits.

So was the break-up of the Bell System all to the good? It was. Due to technical advances (microwave links were the thin edge of the wedge; the advent of optical fibres has widened the breach much further) communications could no longer be regarded as a natural monopoly. The introduction of competition in long-distance services had led to prices lower than they would have otherwise been. To my mind the big loser

[3] It seems to be a contradiction in terms but the American language can put up with practically anything.

was research. A large part of Bell Laboratories was hived off to serve the interests of the Baby Bells. The original, undivided Bell Laboratories represented a quality of research not found anywhere else in the world. Which other industrial laboratory could boast of several Nobel Prize winners? That has now gone. The research laboratory that remained no longer has the resources to ensure its leading position. I suppose other laboratories took up the torch but none of them seems as bright to me.

Privatization in the UK

In the UK the scene was entirely different. Ever since 1911 telecommunications was part of the Post Office monopoly. Apart from times of war, governments, whatever their political hue, interfered little.[4] It was left to the Post Office, the supplying companies, and the trade unions to find a *modus vivendi*. The public was not part of the consultation process. It was taken for granted that members of the public would grumble anyway, so why consult them?

The idea of privatization came mainly for political reasons. Political events in the 1970s gave rise to the theory that labour unrest will continue as long as there are any nationalized companies. The main argument was that nationalized companies could not go bankrupt (the state was always there to bail out the company); hence, trade unions could not lose anything by claiming higher wages—and higher wages in the nationalized industries destabilized private industry too. Furthermore strikes had the tendency to degenerate into political strikes. The Conservative government of Ted Heath was brought down by the miners in 1973, and as a matter of fact, trade unions were largely responsible for the fall of Jim Callaghan's government in 1979 as well.

There was, of course, no shortage of economic arguments in favour of privatization. Nationalized industries were inefficient because overmanning and restrictive practices were widespread,[5] but perhaps the most compelling reason for privatization was that the nationalized industries did not know how much the various services cost. They were greatly involved in cross-subsidization without knowing the extent of the subsidy.

Mrs Thatcher's government came to power in 1979 with a commitment to privatization. The first industry to go was British Petroleum, followed by Cable and Wireless and a number of others. British Telecom (BT) was set up in 1981 as a public corporation separate from the Post Office. Just over half (50.2 per cent) of its stock was privatized in 1984, raising for the government £4 billion. The share price was determined by a combination of political and economic considerations. The economic consideration was to fill up the coffers of the Treasury. The political consideration was to generate popular enthusiasm for the privatization

[4] By lack of interference I mean that the government did not tell the Post Office how to run their business. But there was interference in the sense that the government held the purse strings. A study in the 1980s showed that the fault rate in the telecommunications system correlated nicely with Treasury largesse, or rather the lack of it: less money, more faults.

[5] It may be worth telling a story about a demarcation dispute as I heard it from an old hand who used to work at Standard Telephones and Cables. The company was installing a new electronic exchange alongside an old Strowger one when one of the more vigilant Post Office engineers noticed that the reliability of the Strowger equipment had significantly improved. The only reason he could imagine was that the STC engineers had done some unauthorized repairs, an act clearly violating the inter-trade union agreement. Fortunately, before the dispute could spread beyond the walls of the exchange, the STC engineers offered a plausible explanation for what happened. The installation was a 24-hour operation and the heating was kept on all the time. The warmer, stable temperature made the Strowger equipment work more reliably.

process by ensuring a hefty profit to those buying the shares. The offer price was 130p which jumped to 180p overnight.

To encourage competition was also an aim. Had the government taken it seriously they could have devised a new structure on the American model. But political considerations dictated moving ahead fast. In the time available no more could be done than to plant the seeds of some competitive activities. Mercury plc, a newly established firm, received permission to build a long-distance infrastructure which they did by setting up microwave and optical fibre links. They could, of course, duplicate only a fraction of the BT network. In order to survive Mercury needed interconnection facilities. As part of its licence BT was required to provide them.

In contrast to deregulation in the US, it was clear that the competition clauses hardly affected BT's dominant position. It had to be acknowledged as a private monopoly and consequently had to be regulated. There was obviously plenty of discussion about how the regulation should be affected. In an interesting proposal Alan Walters, the prime minister's economic adviser, advocated a tax inversely related to output. The more BT expanded, the less tax they were going to pay. It would have had the merit of encouraging BT not to rest on their laurels. The regulatory framework finally adopted was due to Stephen Littlechild of Birmingham University. BT had to comply with the requirement that a basket of its prices could not rise faster than RPI – X, where RPI stands for the retail price index and X was set at 3 per cent for a period of 5 years.[6] This meant that prices in real terms had to decline year by year.

To rely on one single formula is a rather blunt weapon, but it worked. The savings came mainly on long-distance calls where margins were wide. Thus, the main beneficiaries were business users. After the expiry of the 5-year period in 1989 the agreement was renewed for a further 4 years with X increasing to 4.5 per cent. Conditions became even more stringent in the 1993 review, which raised X to 7.5 per cent. Naturally, BT could not go on indefinitely to lower their prices. The RPI – X formula is dead by now (March 2020). Not exactly dead, because there is still talk about keeping down prices but the duties of the regulatory authority OFCOM (standing for Office of Communications founded in 2003) have been multiplied and became much vaguer. It is supposed to help both business and consumers—not an easy job. Its duties are well summarized by the icons on the front page of their annual report. OFCOM is responsible for the well-being of an amazing range of facilities (mobile and home phones, broadband, fibre communications, broadcasting). It is admitted in their report that the new online opportunities offer benefits but also harms at the same time. They promise that they will continue to deepen their understanding of the harms and the steps they might take to help address them.[7] It is not clear at the moment what will happen to the BBC but there is no doubt that the Conservative government formed

[6] The complication of the formula comes in the contents of the basket. In the first round only about 50% of BT's total sales were included. International telephone services, leased lines, apparatus, and connection charges for new lines were explicitly excluded.

[7] This language reminds me of Giraudoux's play *La Folle de Chaillot* in which one of the principal characters exclaims: 'I like everything that is beautiful, I detest everything that is ugly.'

after the election of December 2019 wants to clip the political wings of the BBC.

Privatization in France

Nationalization and denationalization have been a political football in France. As late as the 1980s the socialists, heady with their victories in both the presidential and parliamentary elections, were in favour of further nationalization. Soon after acquiring power they decided to nationalize the banks, so, quite clearly, they were not to follow Mrs Thatcher in privatizing telecommunications. But then, in the 1986 elections, the socialists lost. The attitude towards privatization changed but the new conservative government could only take one small step at a time. The first assault on the monopoly position of the DGT (Direction General des Telecommunications) was made in 1986 by permitting competition in setting up cable networks. Two years later mobile telephony was also opened for competition. Had the conservatives stayed in power, further liberalization would have surely followed. But they lost the election in 1990.

This time the change of government meant less radical change. The incoming socialist administration learnt its lessons from the mistakes it had committed in the early 1980s. They were no longer strong advocates of nationalization. In fact, the report by Marc Dandelot, which come out in July 1993, proposed converting France Telecom into a limited company. Not surprisingly, there was strong trade union opposition culminating in strikes. The government caved in.

New elections came in 1994 and with a swing of the pendulum, the conservatives won. Now they proposed turning France Telecom into a private company and selling 49 per cent of its stock.[8] The socialist opposition voted against it in vain. On 31 December 1996 France Telecom turned into a limited company. But then the conservatives became too confident. Parliament was dissolved before their mandate ran out. During the election campaign in the spring of 1997 the government gave the date for selling shares as 24 June. And did they sell the shares? No, the pendulum swung again. The socialists came in and felt that they could not go ahead with the flotation. They knew, however, that they could not stem the tide. An ex-minister, Michel Delabarre, was given the job of looking into all aspects of the problem. He proposed privatization and the opposition to it fizzled out. Proceedings started in October 1997 to sell 20 per cent of France Telecom's capital. Going, going, gone.

It is an interesting story with a lesson. In the climate of the 1990s the battle between ideological purity and economic efficiency was bound to be won by the latter.

[8] Note it is 49 and not 50.2 per cent, as it was in the UK. There is a reluctance even by the French right to relinquish state control completely.

The End of Monopolies?

Once more in the history of communications the Anglo-Saxons led and the rest followed. Governments all over the world have been in the process of giving up their monopolies and permitting the entry of competition, even the entry of foreign firms. Everything is in a flux so there is no point in describing the responses of individual countries, but it is certainly worth noting that nearly all countries within the European Union (Greece, Ireland, and Portugal have been given a temporary reprieve) were required to dismantle their communications monopolies starting in January 1998. In fact, according to the agreement signed by 68 countries in February 1997, they were to liberalize their telecommunications before 2003.

Having said that most of the world is abandoning monopolies, does that include the local networks as well? That's the way things seem to be going. In the US since the Telecommunications Act of 1996 local operating companies have been entitled to compete in the long-distance business provided the regulators are satisfied that they themselves allow competition in the local market. Why this change of attitude? Why is it now believed that local markets no longer constitute natural monopolies? The reason is the advent of mobile communications. If the local market can be served more cheaply by operators using radio waves, then surely that is in the interest of the consumer, and the whole local market needs to be deregulated.

Mobile Communications: The Beginning

Introduction

The desire to communicate between any two points which may move relative to each other must be as old as civilization itself. I have already mentioned in Chapter 1, 'Introduction', a tribal chief who would have loved to receive reports from reconnaissance parties. And the same views must have been shared by all and sundry. Had mobile communications existed it would have been a great benefit to all. Take, for example, the young couple from Verona whose parents were rather antagonistic about the idea of their marriage. Had Friar Lawrence been able to phone the banished Romeo in Mantua, the impending tragedy would surely have been averted.

I suppose that waving a torch while riding a horse was some kind of mobile communications, and the very idea of a lighthouse implies communications between a moving ship and a fixed shore station but, on the whole, it would not be very rewarding to talk of mobile communications before the birth of radio. With the birth of radio it became a practical possibility. The first attempt was made as early as 1898 a mere two years after Marconi's arrival in England. The motivation was not so much to communicate between two persons but rather to show how far away a mobile receiver could pick up signals from a stationary transmitter. The experiments were conducted by the young inventor himself, aided by a few technicians. The transmitter was at Haven Hotel, Poole. The receiver roaming the roads was somewhat bulkier than those in use today, as may be seen in Fig. 14.1.

In the beginning the main application of radio was for communications between ships and shore stations followed by communications between ships. It was extended, in some cases, to communications between trains and fixed railway stations. It was the Lackawanna and Western Railroad in the US which first availed itself of that opportunity. Four fixed stations using long waves were set up at Hoboken, Scranton, Binghampton, and Buffalo. The radio sets on the trains were installed in small cabins (see Fig. 14.2). Operations started in November 1913. A notable success was scored by the radio operators on the occasion of the 'vacation special', which carried 550 students from Cornell University to their homes in New York. They transmitted 128 messages from the students to relatives without a single failure.

Those on the road had perhaps had even greater need for mobile communications. It was introduced for taxis at an early stage because

Getting the Message: A History of Communications. Second Edition. Laszlo Solymar, Oxford University Press (2021).
© Laszlo Solymar. DOI: 10.1093/oso/9780198863007.003.0014

Fig. 14.1 Marconi's mobile receiver.

their journeys were nearly always short, and they had to know where to pick up the next passenger. Lorries were also provided with radio sets because anxious owners wanted to follow their progress as they delivered goods to customers far away. In fact, mobile communications with lorries along the Chicago–St Louis route was MCI's main business before it embarked on its giant-killing exercise.

The main point is, of course, that in contrast to wired services, radio transmitters working at the same frequency will interfere with each other. Hence within a certain area only a limited number of channels

Fig. 14.2 Wireless telephony on the train. The installation is in a small cabin about 2 by 4 feet at one end of the car.

can be available at any time. Nonetheless, if a member of the general public was very keen to acquire a mobile phone and was rich enough to pay for it, then the person could always join a local network, but nothing like country-wide coverage was available.

If the aim is only a private conversation between two people who are some distance from each other, and like to talk while walking, then the communication device to use is a walkie-talkie. It was developed during the Second World War by Motorola for military purposes. The original model needed a backpack to carry; it is very light nowadays to the delight of radio amateurs, and children who can use them as toys. They should, of course, qualify as mobile phones, and that's what they are but they represent merely a sideshow of little interest.

Let's Design a Cellular System

The reasons for mobile communications appearing rather late were purely economic. There was no point setting up a rather elaborate system without a good chance of recouping the investment. So the companies had to wait until the equipment had become cheap enough and the expected demand high enough.

How would we design a mobile communications system if we started from scratch? First, we would choose a frequency high enough so there are enough channels.[1] For the modulation system we would choose

[1] Having previously persuaded the regulatory authorities to allocate to us the required band.

frequency modulation or pulse code modulation since protection against noise is a primary consideration. We would choose a base station and a certain area around it in which communications must be within a certain frequency band. What about power requirements? There we would have some design problems because the power radiated out by the mobile should be high enough to give good reception at the base station, but should be low enough so that the same frequency could be reused a certain distance away. The whole area within which we would want to communicate would then be divided into smaller units. We could call them cells.

What shapes should the cells have? The logical thing is to draw a circle with the base station at the centre. How large should the radius be? Well that would mainly depend on the density of the expected traffic. In our first design-in-principle we should not commit ourselves to more than saying that it should be small in the cities and larger in the country. Naturally, a different frequency band is allocated to each cell so that there should not be interference between adjacent cells. Where can the same frequency be reused? Presumably, some distance away. So it would make sense to allocate all the available frequency bands within a bigger unit which we might call a cluster. The next cluster would use the same frequency bands. A convenient shape for a cluster would be a circle containing seven cells but it looks neater if we represent the cells by hexagons as shown in Fig. 14.3a. Three adjacent clusters may be seen in Fig. 14.3b where the numbers 1 to 7 refer to the frequency bands used.

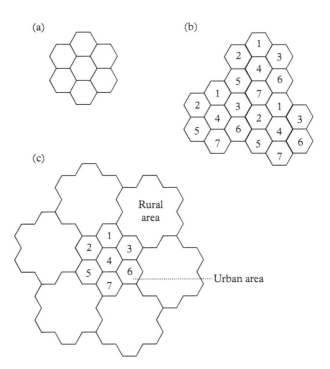

Fig. 14.3 (a) A seven-cell cluster, (b) three adjoining clusters, and (c) clusters of unequal area.

The cells using the same bands would be farthest from each other. A more extensive cellular network might take the form shown in Fig. 14.3c. It consists of an urban area in the centre, and of less densely populated rural areas at the edges.

My main aim so far has been to show that the basic principles of setting up a mobile communications system are not far removed from common sense. The operational principles are fairly simple too. They would mimic the public telephone system with a few features added to account for the roaming of the receivers.

Just as in the public telephone system, a bigger unit must have an exchange which is called a mobile-service switching centre or simply a mobile switching centre. A call from a mobile phone goes first by radio waves to the cell's base station and from there by wire to the cluster's switching centre, which is connected by wire both to the public telephone network and to all the other mobile switching centres. How would the mobile know what frequency to use? It needs to be told. As soon as the mobile enters a new cell it searches for some control signals. Once they are found, the mobile knows where it is and what frequency to use if it wants to initiate a call.

What will happen in the opposite case when somebody from, say, Timbuctoo wants to talk to a mobile phone registered at Lower Edmonton? If the mobile happens to be at its home base then it is quite straightforward. The call from Timbuctoo goes to the nearest base station from where it is transmitted by radio to the mobile's antenna. But what is the procedure if the mobile happens to be in Santa Barbara on the Pacific Coast? The mobile when it entered a cell at Santa Barbara identified itself to the base station as a visitor. Once the base station obtained this information it speedily transmitted it to the mobile's home base at Lower Edmonton. Thus when the call from Timbuctoo arrives at Lower Edmonton it is redirected from there to the right cell in Santa Barbara and the contact is established.

How to Modulate?

The cellular system introduced in the previous section allocates a pair of frequencies (one for base-to-mobile and one for mobile-to-base communications) to each telephone call. This is the 'logical' way of starting up a new system. It is logical in the sense that such systems have amply proven their worth in the past. After all this is how our broadcasting system works. We select a station by turning a dial, i.e. by tuning the receiver to the right frequency.

Changing from an analogue to a digital system, it is still possible to preserve the division by frequency and, at the same time, to introduce time division multiple access, which, within a certain frequency band, allocates time slots to the mobile phones in the cell (similar to the technique described in Chapter 11 when additional channels were inserted between samples).

Another modulation method bears the name code division multiple access. It is now fast gaining ground in the US. It is based on direct sequence, also discussed in Chapter 11. Its essential feature is that a 1 is not simply a pulse and a 0 is not simply the lack of a pulse. Both binary signals are represented by long codes which makes it possible for each mobile to have its own code. The base station is then able to tell which mobile is sending the signal by recognizing its code. Note that long sequences can ensure a large number of possible codes. For example, with a length of 4 units there are $2^4 = 2 \times 2 \times 2 \times 2 = 16$ independent codes. The general rule is that with a code of n units the number of possible codes is 2^n which, with $n = 20$ amounts to about a million. So there is no shortage of personalized codes. With 4 units the possible codes are

1111	1011	0111	0011
1110	1010	0110	0010
1101	1001	0101	0001
1100	1000	0100	0000

Finally, I wish to mention spread spectrum technology, which is also a kind of coding. Its first claim to fame is that it was patented by an unlikely pair of inventors. The patent was taken out in 1941 by Hollywood actress Hedy Lamarr and the composer George Antheil.[2] They proposed that the transmitter and the receiver rapidly change frequencies in a pre-agreed manner (and they had a scheme for how synchronization of frequencies could be achieved) so that the communication would be secure against enemy interception. They offered the patent to the US government free of charge to help the war effort. As far as mobile telephony is concerned, its main advantage is that it can work at very low power. The receiver, knowing the sequence of frequencies, can distinguish it from noise even when the signal level is comparable with the noise level. It seems likely that this method will be more widely used in the future because low power requirement (i.e. very little battery consumption) is a desirable feature for mobile communications.

The Spread of Mobile Telephony

The first mobile phone, a forerunner of the species we use nowadays, was produced by the Motorola company. The first call by that phone was made by Martin Cooper in 1973. It weighed 1.1 kg. It took 10 hours to charge it. Then things started to move. The principles of operation were comprehensively described in 1979 in a series of papers published in the technical journal of AT&T. The first commercial mobile phone was

[2] Born as Hedwig Kiesler in Vienna, Hedy Lamarr was one of the most beautiful women ever to appear in film and she had a head for technology too. She rose to fame in 1933 in the Czech film *Ekstase* on account of appearing in the nude in one of the scenes. She moved to Hollywood before the war broke out, where she played in various glamorous roles until the end of the 1950s.

produced by Motorola in 1983. It cost $3,995. They were obviously designed for businessmen who drove Rolls Royces and were fond of new gadgets. With the passage of time (remember Moore's Law: the number of components that can be put on a single chip tends to double every year), the phones became lighter, less expensive, and the quality of voice kept on improving. The truly mobile telephone had arrived. At the end of the twentieth century there were already over 50 million mobile phones in the US alone.

Cellular telephony was introduced in the UK in 1985, first using an analogue technique. Two companies, Cellnet and Vodafone, were licensed to use a frequency band around 900 MHz. Similar systems came on the Continent as well but each one of them had a different standard. Hence, mobile telephones had limited mobility. They could only be used within a country. A further disadvantage was the complete lack of privacy. Anyone having a suitable receiver could tune in to any conversation. The public's attention was called to this when a telephone conversation of Prince Charles's was taped.

Having recognized the disadvantages of the existing systems the countries of Europe formed a study group in 1982 to develop a new standardized system. It was called Groupe Special Mobile. After lengthy considerations, they decided to introduce a new digital system which became known by the acronym GSM, which by then was anglicized, standing for Global System for Mobile Communications. There was a generous allocation of frequency bands from 890 to 915 MHz and from 935 to 960 MHz for the up- and down-link, respectively. The new system has offered security against unauthorized intrusion, better sound quality, cheaper sets, and wider access. It has been, without doubt, a success story. According to Mike Short by 1999 Cellnet had been deployed in more countries than McDonald's. Launched in 1991 the number of subscribers from 1994 to 1997 increased steeply, as shown in Fig. 14.4.

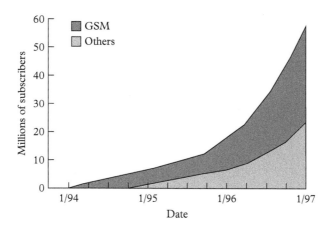

Fig. 14.4 Mobile subscribers worldwide from January 1994 to January 1997.

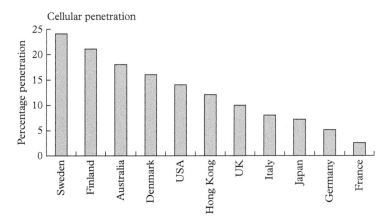

Cellular penetration

Fig. 14.5 Percentage of mobile phones compared with fixed phones.

As with every new technique, some countries are keener than others to implement it. The so-called 'penetration', i.e. the percentage of mobile phones to fixed phones in various countries at the end of 1995, is shown in Fig. 14.5. The Scandinavian countries were leading at the time, and France was trailing at the bottom. The main reason for the French being so much behind was their wavering about privatization and the introduction of competition.

Personal Communications Services

If sound can be transmitted by wireless from person to person in an urban jungle, then why not fax or video as well? Can wireless do everything that can be done with wires? Nearly. It is a question of bandwidth. There is bound to be a scarcity of bandwidth within a given area, so the amount of bandwidth ever to become available by wireless must be rather limited relative to what optical fibres can offer, but it might be enough for most of the services envisaged at the moment.

In order to allow the transmission of more information, a new band around 1800 MHz was allocated to personal communications services. In the UK the system is called DCS 1800, a close relative of GSM, and was introduced by two companies, Orange and One-2-One. The situation is much more complex in the US where there are still a number of competing technologies. Interest is strong, as may be judged from the outcome of frequency band auctions by the Federal Communications Commission. Last time more than 100 hopeful operators were willing to spend around two and a half billion dollars.

The Counter-attack of the Telephone Companies

Local telephone companies have for a long time been able to offer an analogue voice channel and nothing else, and that's clearly not enough for a subscriber today. Their livelihood is threatened from two different directions: cable companies introducing wide band services in the form of coaxial lines or optical fibres, and radio, which may be fixed or cellular.

So what did telephone companies do to catch up with the modern world? Firstly, they invented modems capable of turning digital signals into analogue ones, which can then be sent down an ordinary telephone line. Secondly, they offered new lines capable of transmitting much wider bandwidth. In the UK for example ISDN30, part of the Integrated Services Digital Network, is offering a transmission rate of 2 megabit per second. Also, they perfected digital processing techniques to the extent that they can deliver up to several megabits per second using ordinary copper wire.

Global Mobile Communications by Satellite

I have already mentioned satellites in Chapter 9 on microwaves. Indeed, it was the advent of microwave sources that made possible communications between two stations on Earth via artificial satellites. Microwaves were needed to ensure that the beam will be narrow enough to hit the satellite with sufficient power.

How can we achieve mobile communications? How does a call go from any point on Earth to any other point on Earth? A necessary condition is, of course, that there should be satellite coverage for every point on Earth. The only satellite system that can do that, from pole to pole, is that of Iridium. It is achieved by having the satellites in the right orbits and by communications between satellites. So how do satellites of Iridium help in directing a call from address A to Address B? Let us see the rather tortuous ways signals travel from A to B. It depends, of course, on whether the call is originating by a mobile or landline phone and whether the addressee is a mobile or a landline. Let's look at all four possibilities:

Landline to landline

  (i) From address A by landline to the nearest gateway

 (ii) From gateway to satellite by wireless

(iii) By wireless from satellite to satellite nearest to address B

 (iv) From satellite to the nearest gateway

  (v) From gateway by landline to address B

Landline to mobile

  (i) From address A by landline to the nearest gateway

   (ii) From gateway to satellite by wireless

  (iii) From satellite to satellite by wireless

  (iv) From satellite to gateway by wireless

   (v) By landline from gateway to ground station nearest to address B

  (vi) By wireless from ground station to address B

Mobile to mobile

   (i) From address A by wireless to the nearest ground station

  (ii) From ground station to gateway by landline

  (iii) From gateway to satellite by wireless

  (iv) From satellite to satellite by wireless

   (v) From satellite to gateway by wireless

  (vi) By landline from gateway to ground station nearest to address B

 (vii) By wireless from ground station to address B

Mobile to Landline

   (i) From address A by wireless to the nearest ground station

  (ii) From ground station to gateway by landline

  (iii) From gateway to satellite by wireless

  (iv) From satellite to satellite by wireless

   (v) From satellite to gateway by wireless

  (vi) From gateway by landline to address B

We have exhausted all four possibilities. In each case there was some reliance on existing networks. What if we are at a construction site somewhere in the wilderness and there are no ground stations nearby? For this exercise we can further assume that not only address A is in the wilderness but address B as well. What then? In fact, this is the simplest case so far. The call goes directly to the nearest available satellite, then from satellite to satellite until one of the satellites is near to address B. Then again the call goes directly to address B. There is, of course, a price to pay when this is the only solution. The satellite phones need more power and when in receiving mode it needs to be highly sensitive. Not surprisingly, phones that can directly address satellites are heavier and cost about five times as much as an ordinary mobile phone.

An example of how efficient global mobile communications are can be seen in Fig. 14.6.

Fig. 14.6 Coverage becomes worldwide.

The Human Factor

For whom were mobile communications originally set up? Clearly for businessmen, for people who had the money to pay for it and who regarded themselves as terribly important. A new mobile phone went nicely with the new Mercedes. As prices declined, more and more people managed to convince themselves that it would be nice to have access to a telephone at all times.

The clear economic consequence is a straight transfer of money from consumers' pockets to the coffers of the telephone companies. A typical example is Orange in the UK. Its shares were floated in the spring of 1996. Their price doubled in two years.

The social advantages are obvious. Better communications is obviously a good thing. One major disadvantage is the resulting loss of privacy. All mobile subscribers are continuously exposed to calls inflicted upon them by friends, enemies, and business contacts.

A further disadvantage is that one's whereabouts are given away. It is therefore advisable for anyone intent on committing a major crime to have their mobile switched off. The same applies to rebellions. Dudayev, the Chechen rebel leader, was careless enough to use a mobile phone in an exposed position. A Russian rocket hit him there.

Acronyms

I am not a man for crusades. I feel most causes can get on without me. I would, however, join a Campaign for Curbing the Spread of Acronyms at any time. Why do I make this point in the chapter on mobile communications? Are they the worst offenders? I think so. In a recently published book on the subject I counted not less than 300 acronyms.

Some of them were really horrid. Take, for example, FPLMTS (Future Public Land Mobile Telecommunications System) or GMPCS (Global Mobile Personal Communications by Satellite). Could anything be worse than that?

If I had my way I would set up an Acronym Approval Agency. As a minimum requirement acronyms seeking approval would have to be easily memorable and, preferably, would give some intimation of what the whole thing is about. As examples I give here three acronyms which I find least objectionable: POTS (plain old telephone system), FLAG (Fibre-Optic Link Around the Globe), and SWIFT (Society for World Interbank Financial Telecommunications).

The Fax Machine

The Basic Principles

Fax, as most people will know, is a contracted form of facsimile. It comes from two Latin words *facere* = to do (the imperative form is 'fac') and *simile* = similar. According to the *Oxford English Dictionary* the first use of the word appears in 1691. The sentence quoted in the dictionary is 'a fac simile might easily be taken', meaning by it an exact copy. In this sense taking a facsimile goes back a long time. The motivation to take an exact copy must be as old as writing itself. It seems fairly likely that as soon as Moses came down from the Mount of Sinai (and the little incident with the golden calf was over) several craftsmen must have been busy taking a facsimile of the stone tablets.

The present chapter will, of course, be concerned with taking a facsimile in a very specific sense, namely with the process of sending an exact copy from point A to point B. How can one do it? Can it be done in the same way as sending a telegraph? Can it be done with a mechanical telegraph? Did anyone ever send an exact copy of a picture by a mechanical telegraph? The answer is no, but, in principle, it could have been done. A very simple semaphore with five possible positions would have been able to accomplish the task.

Let us assume that Napoleon, looking fondly at one of the portraits painted of him (Fig. 15.1a), suddenly decides that he wants the rest of the nation to share the pleasure. It could have been done in the following manner. Firstly, the portrait is divided up into (say) 10,000 small squares by drawing 100 vertical and 100 horizontal lines. If, within a square, there is more black than white (the judgment would have been delivered by a specially trained clerk), then the square would be declared black. In the opposite case it would be declared white. After performing that operation for each square the portrait would have looked like that shown in Fig. 15.1b, still a reasonable image. In order to send it via a chain of semaphores all that needs to be done now is to specify whether a particular square is white or black.

The transmission of the picture may take the following form. First, the next tower needs to be told that transmission starts. That might be coded by the signal shown in Fig. 15.2a. Next the information is sent that the first square is black. For that, the code shown in Fig. 15.2b may be chosen. Having submitted the first piece of information the semaphore may now return to its original position (Fig. 15.2c). Then comes the turn of the next square. If it is black the same code goes up again. If it is white it will

Getting the Message: A History of Communications. Second Edition. Laszlo Solymar, Oxford University Press (2021).
© Laszlo Solymar. DOI: 10.1093/oso/9780198863007.003.0015

Fig. 15.1 (a) A portrait of Napoleon, (b) portrait replaced by ten thousand black-and-white squares, and (c) portrait replaced by forty thousand black-and-white squares.

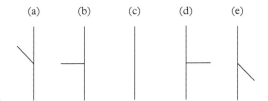

Fig. 15.2 A possible solution for sending a fax by semaphore.

be coded by that shown in Fig. 15.2d, after which the semaphore returns again to its rest position. Thus, two distinct positions of the semaphore will give information about the state (black or white) of one particular square. For 10,000 squares 20,002 signals are needed altogether, including one (say in the form of Fig. 15.2e), indicating the end of the message.

Assuming that each signal can be given in five seconds, a picture of the quality of Fig. 15.1b can be transmitted in about 28 hours. At the receiving side a similar mesh would have to be ready, and whenever a 'black' signals is sent a clerk sitting there would nicely blacken the corresponding square. The picture quality at the receiving end could be improved by increasing the number of vertical and horizontal lines to (say) two hundred, resulting in the improved version of Fig. 15.1c, but the transmission time would increase to 112 hours. To send information consecutively about the state of each square is, of course, only one of the possible codes. In the court of Napoleon there would have been plenty of mathematicians to point out that much faster codes exist. One could, for example, give the number of black squares followed by the number of white squares, and then again the number of black squares, etc.

Early Attempts

The chances of sending a picture by telegraph very much improved with the invention of the electrical telegraph. It would have been, for example, possible to code a white square with a dot and a black square

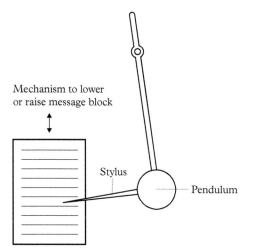

Mechanism to lower
or raise message block

Stylus

Pendulum

Fig. 15.3 Schematic representation of the fax machine invented by Alexander Bain.

with a dash. Assuming a reasonable speed of 8 signals per second, the transmission time for our lower quality picture would have come down to about 40 minutes which might have proved acceptable. So did fax transmission start soon after the invention of the telegraph? It did not, but it started much sooner than most people think. I did actually pose this question to a number of my colleagues in the following form: 'When and where, do you think, was the first commercial service introduced capable of transmitting facsimile (that is text plus pictures) by electrical means?' The earliest estimate I received was 1924. Most of the answers put the beginning of the service between 1960 and 1970. As for the venue, most people expected that it was first introduced in the US, although one thought that Japan had the priority. The correct answer is 1865. The two cities connected were Paris and Lyons.

The idea was born in 1842. It came from Alexander Bain.[1] He took out a patent in the following year. The principles of operation of Bain's fax machine may be understood from Fig. 15.3. The letters to be transmitted had to be made out of metal. A stylus attached to the pendulum swung past the message block. Whenever it touched a raised letter an electric circuit was established. The message block was lowered by one notch after each sweep of the pendulum, enabling the whole message to be read. At the receiving end the movement of an identical pendulum was synchronized with that of the transmitting pendulum and an identical stylus scanned a chemically treated paper.[2] The paper changed colour whenever an electric current flowed. The transmitted document appeared as blue on a white background. The idea was good but there is no evidence that the device was ever built.

A similar idea with a stylus and synchronized movement was due to Frederick Bakewell around 1850. A cylindrical metal sheet was fixed to a drum which could rotate round its axis. The message was written on the sheet with non-conducting ink. As the drum rotated a screw mechanism

[1] There is an entry under 'Bain, Alexander' in most encyclopaedias. He was a Scotsman who lived in the nineteenth century. Alas, the Alexander Bain of the encyclopaedias was a philosopher, a man who happened to have the same name and happened to live at about the same time as our hero. The difficulties of rich men trying to enter the Kingdom of Heaven are well known. Technologists trying to gain entry in encyclopaedias do not seem to fare much better.
[2] Synchronizing clocks by electrical signals was also one of Bain's inventions but that wasn't judged to be of sufficient importance either for inclusion in encylopaedias.

drove the stylus forward. The electrical circuit was interrupted whenever the stylus touched the ink. Reception was by a similar device, again
using chemically treated paper. The main problem was synchronization
between the movement of the transmitting and receiving stylus. The
device never worked satisfactorily although a sample of a transmitted
message is reputed to have been shown at the Great Exhibition of 1851.

The next man to take up the challenge was Giovanni Caselli, an
Italian priest who had equal interests in religion, science, and politics.
Involvement with the last-mentioned pursuit forced him to seek asylum
in Florence in 1849. He resurrected Bain's pendulum idea and combined
it with Bakewell's insulating ink. Finding no interest in Italy he went to
Paris in 1857, where he secured the help of Gustave Froment, the leading
French instrument-maker of the time. By 1860 they had managed to
complete two machines. The first demonstration took place on 10 May
1860 in the presence of Napoleon III (Fig. 15.4), who was always interested in technical novelties. The message to be transmitted between the
two machines contained the text, 'God bless the Emperor, God save the

L'Empereur visite les bureaux du télégraphe électrique au Ministère de l'Intérieur.

Fig. 15.4 Napoleon III inspects Caselli's fax machine at the Ministry of the Interior.

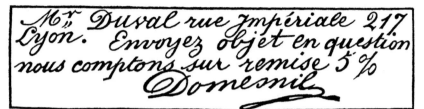

(a)

(b)

Fig. 15.5 Faxes sent by Caselli's machine, (a) in French ('Send object agreed. We count on 5 per cent discount'), and (b) in Chinese ('Confucius said: He who travels much will see much and will learn much. He who climbs high will realize how small he is').

Emperor for the glory of France, the liberation of Italy and the welfare of the whole world', decorated with the Imperial eagle. The experiment was successful. The emperor could not fail to be impressed. He authorized Caselli to use the Paris–Amiens telegraph line for further experiments.

In September 1861 Victor Emmanuel (the king of recently united Italy) invited Caselli to demonstrate the fax machine at a Florence exhibition. The young Prince Amadeo could not contain his excitement as a copy came off the receiving machine. 'It's daddy, it's daddy, it's daddy's portrait', he exclaimed.[3]

In 1863 the Conseil d'Etat authorized the setting up of facsimile communications. Regular service started in 1865 between Lyons and Paris, extended to Marseille in 1867. By that time there were four machines working on the Paris–Lyons line. The peak performance was a quite respectable 40 words per minute. Unfortunately, there was no great demand for Caselli's faxes. To send a telegram was much easier and much cheaper. Fax had, of course, the advantage of being able to transmit Arabic and Chinese scripts. Photographs of such original faxes may be seen in Fig. 15.5. The number of faxes sent amounted to no more than about a dozen a day, surely not an economic proposition. So why were fax communications kept alive for five long years? I suppose for the same reason that it was first installed: to show the marvels of engineering. In Britain's commercial atmosphere such a venture would have never got off the ground. In France the government was willing to support prestigious projects, even if they were run at a loss. The fax service using Caselli's machines came to an end with the Franco-Prussian war. It was not resurrected after the war.

Image transmission using the photoelectric properties of selenium (i.e. it becomes conductive in the presence of incident light) was carried out by George Carey in Boston in the middle of the 1870s and by Noah Amstutz in Cleveland at the beginning of the 1890s, but neither was a commercial success. Americans were not as yet ready for the fax.

The fax machine that was a commercial success was invented by Arthur Korn of Germany in 1902. It appeared just at the right time when photography was becoming widespread and cheap, and illustrated newspapers competed with each other to bring the latest news in pictorial form. One of the pioneers was the French journal *l'Illustration*. It announced the new invention in its issue of 24 November 1906, the front page of which shows a photograph transmitted from Berlin to Paris (Fig. 15.6). Korn combined in his machine Bakewell's rotating drum and Amstutz's selenium cell. The stylus was replaced by a narrow beam of light which could cross the transparent cylinder to fall upon the selenium cell inside. As a result an electric current flowed except when the incident light was blocked by non-transparent parts of the message. The information was contained in the variation of current which could reproduce the original picture at the receiving end.

A Frenchman, Edouard Belin, had similar ideas but was beaten by Korn at the patent office. Undeterred, he kept on working and in 1913 managed to produce the world's first portable fax machine (Fig. 15.7). It enabled more muscular reporters to carry the case with them, take photographs, develop them, and send them immediately to newspaper offices by plugging the machine into an ordinary telephone line. These portable fax machines were used in the First World War for sending photographs from the front lines.

Pictures were first transmitted by telegraph and telephone lines but there was, of course, no reason why the same information could not travel via the courtesy of radio waves. H. Knudsen, who built his own machine, demonstrated such a possibility in London in 1908. The French newspaper *Le Matin* managed to transmit pictures of acceptable quality across the Atlantic using Belin's portable machine in 1921. Korn was not much behind. He succeeded a year later in sending a photograph (that of Pope Pius XI) by radio from Rome to Bar Harbour, Maine.

The Americans woke up rather late to the commercial potential of picture transmission. Why? Possibly because American newspapers served a local population with a limited interest in the world outside. Maybe the advent of the radio and the possibility of receiving photographs and newspaper articles from Europe (from the mother country whether it was England, Ireland, Scotland, Italy, or Sweden) tipped the balance. According to one account the chairman of the board of directors of the Radio Corporation of America (RCA for short) complained in an after-dinner speech that the news from Europe was still jotted down by telegraph clerks one letter at a time. He said:

Fig. 15.6 Fax sent from Berlin to Paris using Arthur Korn's fax machine in 1906. The original photograph is in the lower right-hand corner.

the new possibilities of radio should make it feasible for us to say: 'ZIP, and a page of the *London Times* is in New York City'. 'Not being an engineer', he added, 'I am not interested in the details; that is your job.'

The engineers did indeed get cracking. In 1924 RCA completed the feat of sending a picture across the Atlantic and back. It travelled from New York by wire to New Brunswick, New Jersey; from there by radio to Brentwood, England; then by wire to London whence by wire again to Caernarvon in Wales; then by radio to Riverhead, Long Island; and

Fig. 15.7 Belin's portable fax machine.

Fig. 15.8 First photoradiogram: New York to London, back to New York, 6 July 1924.

finally back to New York. The quality of the picture was still quite good as may be seen in Fig. 15.8. By 1926 RCA introduced a commercial fax service a year behind their rival, AT&T. From then on, faxes travelling across the Atlantic were a matter of routine. The reason for the rather infrequent use was, first, the price and, second, the lack of an accepted standard. Fax machines that could talk to each other had to come from the same manufacturer.

Faxing Newspapers

In 1937 a newspaper was delivered by fax via radio waves. The idea was simple. Printing newspapers in great numbers was no longer necessary. As an adjunct of radio programmes, radio stations could also broadcast facsimile editions of newspapers which the receiving machines in the home of each subscriber would print out.

Further attempts in this direction stopped when the Second World War broke out. There were more urgent things to consider. The idea was, however, revived after the war. Four major newspapers, *Miami Herald*, *Chicago Tribune*, *Philadelphia Inquirer*, and *New York Times*, all had experimental fax transmissions. A photograph of children receiving an experimental transmission of comics by fax is shown in Fig. 15.9. The

Fig. 15.9 Children watching as a fax machine churns out a comic.

idea might have come off if television technology had had another decade of teething problems. As it happened the availability of television killed the project before it was born. People preferred to look at moving images.[4]

Maturity

Could the fax machine have died in its infancy, never to reach maturity? It could have happened. By the 1970s, paper had a bad press. The idea of a paperless, electronic office was in the air. To introduce a new device churning out endless reams of paper ran against conventional wisdom. So how did the fax become accepted?

The main obstacle to general acceptance was the lack of standards. Fax machines designed for the office had existed since the 1950s but they were expensive and the various makes could not be interconnected. The first major step forward was the recommendation for an international standard by the United Nations' International Telecommunications Union in 1968. It became known as Group 1. The machines transmitted an A4 page in 4 or 6 minutes. The next advance was due to a combination of Japanese need and Japanese know-how. They needed a fax machine because their two thousand characters could not be transmitted by teleprinters. At the same time they were able to use large-scale integrated circuits to develop a digital version which was both cheaper and better than the existing analogue models. This happened in the early and middle 1970s when digitalization was still a novelty.

The inadequacy of the Group 1 standard was clearly recognized. Group 2 came as a stop-gap measure in 1976 followed in 1980 by Group 3, the present digital standard. The corresponding fax machines scan 800 × 1200 points on an A4 page in 1 minute. This is 960 kilobits per minute, corresponding to 16 kilobits per second. According to the argument used in Chapter 11 (see Fig. 11.5), a bandwidth of 8 kHz is needed to transmit this amount of information—beyond the capabilities of most telephone lines. The solution was partly to find better codes than a 1 for a black square and a 0 for a white square and partly to introduce a new modulation method which relies on both the phase and amplitude of the signal.[5]

A further factor facilitating the mass introduction of fax machines was the Carterfone decision in the United States which forced the telephone companies to allow direct access to their telephone networks. The Post Office monopolies in Japan and in Europe followed suit. After that it was the usual feedback mechanism that ensured the take-off. Lower prices led to more offices adopting the fax machine, and a higher number of machines sold led to economies of scale in the manufacturing process—hence to lower prices. In fact, the price of fax machines declined by a factor of 30 in the period between 1980 and 1992.

[4]There is no difference in principle between fax and television. Only the speed is different. Television offers 25 frames a second. Fax machines can do one frame in a minute.

[5]One such code was mentioned earlier in this chapter: instead of sending information about each pixel, it is the number of white pixels followed by the number of black pixels in a row that is transmitted. A further simple way of reducing the information content is by vertical correlation, i.e. by relying on the fact that two pixels above each other are likely to be both white or both black.

In the UK there was another factor that helped the introduction of fax machines. The country's very efficient postal service ground to a halt in 1987 owing to a strike by postal workers. A mathematical model by Lyons *et al.* of the BT Research Laboratories clearly showed that a suspension of some service even for a short time can lead to the take-off of a rival service.[6]

Group 3 is still with us although a faster standard, called Group 4, was approved in 1984 by the International Telecommunications Union. The difficulty with the introduction of Group 4 was that it was not compatible with Group 3. Thus, a change to Group 4 fax machines would have required some major investment. It has not been very popular.

Privacy and Legal Issues

The fax cover sheet is an attempt to make the communication private. It does not work in practice. Faxes usually hang around in some central office for everyone to read until the recipient picks it up.[7] Privacy could only be ensured if everyone had his/her own fax machine—an unlikely scenario.

Another problem with the technology is that fraud is very simple. All one has to do is to copy a heading on the top of a letter and the signature at the bottom, and write in between a quite different message. Authentication would be possible in principle but is it worth the bother?

Telex versus Fax

As mentioned in Chapter 7 telex service in the UK started well before the Second World War. Its success continued after the war. The number of subscribers increased quite steeply up to the 1980s but then decline set in, as shown in Fig. 15.10. The tendency was similar in France, as shown in the same diagram. The reason for the decline is obviously the spread of the fax machine. The capabilities of fax were far superior to those of telex, and a fax machine, once bought, could simply be plugged into an existing telephone socket. Not surprisingly, telex was completely supplanted by fax.

Will telex completely disappear? Yes, it will. It will obey the rule that all animate and inanimate objects follow. Paraphrasing slightly William Shakespeare we could say that "they strut and fret for a while on the stage and then they are heard no more." The speed of demise in this particular case depends mainly on developing countries with whom many of the business contacts are still conducted by telex.

The number of fax machines installed in the UK and in France is plotted in Fig. 15.11 for the period from 1986 to 1996. For international comparisons it is probably better to relate the number of fax machines

[6] BT Technology, Vol. 6, July–August, pp. 431–7, 1995.

[7] The fact that it is so easy to pick up a fax from an office or even when it comes off the fax machine may lead to far-reaching consequences. I remember a French film, *Le Concert*, where the initial events take place in Moscow in the Bolshoi Theatre before the collapse of the Soviet Union. The night cleaner, who used to be a famous conductor before he defied authority, picks up in the middle of the night a fax inviting the Bolshoi's orchestra to a performance in Paris. The cleaner does not say a word to the director but organizes an orchestra from all those musicians who had been dismissed for political reasons and appears with them in Paris. A comedy, of course, but very entertaining.

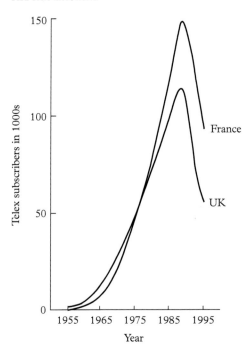

Fig. 15.10 The rise and fall of telex in France and in the UK.

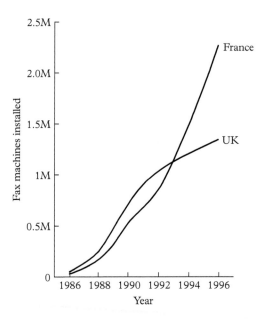

Fig. 15.11 The number of fax machines installed in France and in the UK in the period 1986 to 1996.

installed to the population of the country. In Figs. 15.12a and 15.12b the number of fax machines per 1,000 inhabitants is shown for the year 1996. Japan leading the world is not particularly surprising. Fax is ideal for Japanese characters, as already recognized by Edison. The US being second is due to a population keen to buy a gadget once its price drops quite low. Belgium, I suspect, is in the third place due to all those faxes

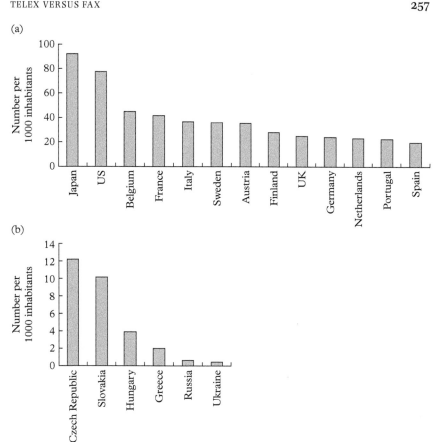

Fig. 15.12 International comparison of fax machines installed in 1996.

to do with European business. France is high up because successive French governments looked at telecommunications with a benevolent eye, willing to put in a bit of subsidy here and there. The fifth place of Italy may be explained by a postal service which occasionally lapses into anarchy. Conversely, the low position of the UK is probably due to a good postal service which makes the fax less important. The low German position is probably explainable by the predicament of East Germany which at the time still bore all the marks of a people's republic. I have no explanation for the low position of the Netherlands and find it rather surprising that Portugal comes out above Spain. Countries in Central Europe, in the Balkans, and in Eastern Europe are considerably below Western Europe (see Fig. 15.12b), but their stock was growing fast at the time.

Jump now to 2020. Is the fax following the telex to oblivion? It is bound to but it does not willingly accept its own demise. Its life has been extended by the invention of efax that makes the actual fax machine superfluous and for transmission uses the Internet instead of telephone lines. There are still about 40 million fax machines operating in the world sending billions of fax messages in a year. Alas, its prospects to survive another decade are minimal. In the UK the National Health Service decided to completely abandon the use of fax machines from 31 March 2020. It is not too early for museums to exhibit a specimen for the public to admire how things were done before the computer revolution.

The Name

Throughout this chapter I have been using quite consistently the modern terms 'fax' and 'fax machine'. Most modern terms are identical with those first introduced (a telephone is still a telephone, a trunk line is still a trunk line, frequency modulation is still frequency modulation). For the fax it is different. It was so often rediscovered and there were such frequent changes in its main applications that it has been described by various names. Starting with the first patent, Alexander Bain called it an 'automatic electro-chemical recording telegraph'. Bakewell thought of it as a 'copying telegraph'; Caselli introduced the terms 'pantelegraph' (*pan* = 'every' in Greek, indicating that it can telegraph everything) and 'autograph'.

The twentieth century was more sober. As the main use was sending photographs great distances, the natural description was 'telephotography' or 'phototelegraphy'. Sometimes it was called after the inventor. The portable machine invented by Belin was referred to as a 'Belinograph' or, by those who got to appreciate it, as a 'Belino'. If the means of transmission was radio waves, the term often employed was 'photoradiogram transmitter' or receiver.

So when was it first called 'facsimile transmission'? The name facsimile already existed in the relevant (that is Italian, French, and English) languages, so that could have been the description from the earliest times, but neither Bain nor Bakewell chose it (perhaps they did not have a sufficiently good classical education). Caselli, a priest, was aware of the term of course. In 1855 in a letter to the l'Institute de France he wrote: 'L'idée de reproduire, par l'électricité et de grand distances, un fac-simile de l'écriture et du dessin',[8] so he used the term but not for naming his machine. The modern French name *telecopieur* does not refer to facsimile either.

In order to find out when the learned authorities of Oxford gave their blessings I went through the copies of the *Concise Oxford Dictionary*, starting with the first edition in 1911. It is not until the sixth edition in 1976 that the old meaning of facsimile is supplemented by a new one: 'system of producing this by radio, etc. transmission of signal from scanning of writing, etc.' A more intelligible (and intelligent) definition appears only in the eighth edition of 1990: 'production of an exact copy of a document, etc. by electronic scanning and transmission of the resulting data', and by then there is also an entry under 'fax'.

As for Caselli, he is remembered by Italian and French encyclopaedias but not by English or American ones.

[8] 'The idea of reproducing, by electricity at a large distance, the fac-simile of writing or of a drawing'.

16

The Communications–Computing Symbiosis: The Beginning

Introduction

Symbiosis means a mutually advantageous association of organisms or persons, the term coming originally from biology. It sounds a bit highfalutin, but it is quite a good way of saying that communications and computing have been helping each other. All modern communications equipment has a computer in it, and most computers have the means to communicate with other computers.

The two subjects used to be entirely separate. When Charles Wheatstone worked on his telegraph and Charles Babbage on his difference engine (the earliest form of a computer), they did not think for a moment that their endeavours would, in the fullness of time, lead to a new, interrelated discipline.

Computers have already made a brief appearance in our narrative when talking of digital exchanges and digital signal processing. In the present chapter the emphasis will be on computing. Firstly, I shall try to tell the extraordinary story of a bunch of teenagers who created a multibillion-dollar industry and became billionaires in the process. Then will come the story of a government anxious to evolve plans for maintaining communications in the wake of a nuclear disaster, and how it led to the unpredicted and unpredictable emergence of the Internet, e-mail, the World Wide Web, and the opportunity for everybody to talk to everybody else using that new medium.

The Rise of the Personal Computer

Computers used to be big. The first computer I ever used was called Stantec Zebra (produced by Standard Telephones and Cables): it occupied a big room. It was personal only in the sense that I got a personal call from the man in charge when it was my turn to use it. It was 1958. There were no long queues standing in front of that impressive-looking machine. Not many in the company showed any interest in learning the code, writing the program, punching it patiently upon a paper tape, feeding it into the machine, and waiting to see what went wrong. Because it never worked the first time.

But then things started to change. The invention of the integrated circuit in 1959 made computers smaller and more reliable. Moore's Law

Getting the Message: A History of Communications. Second Edition. Laszlo Solymar, Oxford University Press (2021).
© Laszlo Solymar. DOI: 10.1093/oso/9780198863007.003.0016

(that the number of components on a silicon chip tends to double every year) was already in operation. Everything tended to become smaller. The 1960s celebrated the advent of the miniskirt, the minicar, and the minicomputer. But minicomputers were neither very small nor powerful enough. Mainframe computers still reigned supreme.[1] Then in 1970 came the invention of microprocessors by Robert Noyce at Intel. The obstacles in the way of developing a small desktop computer were suddenly removed. So it was technically possible to produce them, but was there any need for them? Nobody could be certain. The only consoling thought was that there had not been much clamour for electronic calculators either, but when Texas Instruments brought them out they managed to sell 5 million in 1972, a year after their launch. So did any of the major companies consider developing a small desktop computer? Xerox did. Mostly as an insurance policy. People started to talk about the 'paperless office' and Xerox did not want to be left behind. The New York head office gave the task to their Palo Alto Research Center, known as Xerox PARC, set up in 1971. And lo and behold they developed a computer which fitted upon the top of a desk (admittedly, some of it had to be shoved below the desk).

It was the engineers of Xerox PARC who introduced the most important innovation of the time, the graphical user interface, known today as GUI or simply gooey. The basic idea was that the screen should look like a desk with lots of files heaped upon each other and there should be some facility (it became known as a mouse) to make the files appear when the user wanted to work on them, and make them disappear into the storage area when they were no longer needed. They called their computer Xerox/Alto. It was more or less ready in 1973.

Did Xerox rush into mass production? Well, one of the difficulties was that it was far too expensive. It would have cost tens of thousands of dollars and the market would have been unlikely to pay that much. But that should have merely indicated the need to initiate a crash programme to reduce costs. They did nothing of the kind. The principal reason for their reluctance to proceed was simple old-fashioned fear of taking risks. In contrast to expectations the threat of a paperless office seemed to recede into the uncertain future. So why bother? Xerox head office decided to do nothing.

While Xerox was sitting on its laurels a small computer company with the enthusiastic Ed Roberts at its head decided to bring out a computer kit. It was called Altair and was built for other computer enthusiasts. It had to be assembled from its components and then either it worked or it didn't. In either case it represented a challenge. The customers were mainly teenagers who caught the electronic bug at an early age. By no stretch of the imagination could these computers be regarded as user-friendly. They did not have a keyboard or a monitor or a printer. They had to be programmed by rearranging switches and

[1] 'Mainframe' conjures up the image of something big and important like mainland or mainstream.

the answer was displayed by bulbs lighting up. It was a great toy for enthusiasts.

Then came two young Harvard undergraduates, Paul Allen and Bill Gates (19 at the time), who offered to make the Altair computer more useful by providing it with a computer language of its own. Well, not entirely its own. BASIC was already a well-known computer language,[2] but Paul and Bill adapted it for the Altair computer. Others attached some terminals to the Altair so that it actually became quite useful. One could use it for playing games or even for some rudimentary kind of word processing.

Paul and Bill liked the job. Producing software was fun. They said goodbye to Harvard and set up their company at Albuquerque close to that of Ed Roberts'. They called the company Micro-Soft, dropping the hyphen a little later.

The next advance came again from two electronic wizards, Steve Jobs and Steve Wozniak, both in their teens. Their first attempt at producing an original piece of electronic design was a blue box whose output signals would persuade an ordinary telephone to connect them to any subscriber in the world—free of charge. They could see that manipulating electrical signals was fun, so why not build a whole computer? They bought the components off the shelf and produced a computer by putting all the components on a single circuit board. It still had no keyboard, no monitor, no printer but it was good for showing off to their friends. 'Look what we have done? Pretty, isn't it?' However, the young do-it-yourself enthusiasts had to live on something. They thought they might be able to sell their rudimentary computers so in 1976 they founded a company called Apple (Fig. 16.1), and the fruit of their labour was called Apple I. It cost $666.66. They sold as many as fifty of them.

The next adventure was even more successful. Now they had the idea of packaging the computer. They did some strenuous design work (can one chip do the job that three did before?), they put all the necessary

Fig. 16.1 Five apples that played a significant role in Western civilization. The first one hung in the Garden of Eden, before Eve, under the influence of a snake, decided to have a bite with unfortunate consequences for the whole of mankind. The second apple is a golden one, the one that Paris gave to Aphrodite. The third one was shot by William Tell. The fourth one is Newton's that may or may not have fallen upon his head. The most influential one is the fifth, Steve Jobs' apple, that changed the way we live. Source: Ekaterina Shamonina.

[2] In fact, BASIC, an acronym standing for Beginners All-purpose Symbolic Instruction Code, was developed in 1965 by John Kemeny and Thomas Kurtz.

hardware together, and the finished product looked quite professional. All they needed at this stage was to get some capital to manufacture the computers. Who would give several hundred thousand dollars to two scruffy teenagers? A venture capitalist did. The company managed to sell thousands of these computers under the name of Apple II. Its successor the Apple II+, brought out in 1978, was an even greater success. It cost about $1,200 and was marketed both in the US and in Europe.

By 1979 these small computers represented big business. And then the company had an enormous piece of luck which ensured their further success. Steve Jobs had an invitation to visit Xerox PARC. And there he saw the future: the graphical user interface. He asked for a demonstration and brought along his entire programming team. The local management of Xerox PARC showed them all. Why? Nobody knows. Perhaps out of frustration. 'Here we are with an excellent product', they might have thought. 'Here is something that could beat the world and our headquarters in New York don't give a damn. If these youngsters can bring out something using these principles, good luck to them.' So Steve Wozniak and Steve Jobs developed GUI further and incorporated it in their next product. It was a world beater. Everyone liked the GUI. It was so much easier to handle than any of the earlier operating systems.

And then the founders of Apple had a second piece of luck. Dan Bricklin, a graduate student at Harvard, invented the electronic spreadsheet. He called it VisiCalc. It was the first business application, a tool for financial planning. It made possible the testing of all kinds of hypotheses. What if I increase my capital by 10 per cent? What if interest rates are going up? What if I reduce prices by 5 per cent? What if I hire three more engineers? VisiCalc was able to deliver an answer within minutes that would have meant days of tedious calculations before. Apple computers combined with VisiCalc had real business applications. Sales took off even faster. Apple went public in 1980. Steve Jobs became a multimillionaire. The shares were rather volatile. In 1982 they sank to half of the issue price. Within a year they rose by a factor of 5 to decline again afterwards.

IBM, the biggest producer of mainframe computers, woke up to the challenge in 1980. The management wanted a desktop computer designed. How long would that take? They asked their engineers. From previous experience three or four years seemed a feasible figure. But they did not have the time. So they did something entirely at odds with the company's well-established traditions. Instead of manufacturing all the components themselves they decided to buy them from other manufacturers and put them together, just like their teenage competitors did. That was fine for the hardware. Intel was only too happy when IBM ordered another fifty thousand microprocessors and the manufacturers of keyboards, monitors, etc. also had big smiles on their faces as the orders came in. All the hardware was there to be assembled to the satisfaction of the whole computer industry.

What about software? Up until then software had been a trivial problem. Given enough warning any competent group of programmers could have delivered the necessary software. But IBM was in a hurry. They quickly needed both a computer language to be adapted for their new personal computer (as they became called) and an operating system (which determines how the various pieces of hardware work together). As it happened there was only one suitable operating system on the market which could have been used. That was Gary Kildale's CPM at Digital Research, a small company in Florida. Some of the top brass of IBM went on pilgrimage, visited Digital Research, and asked humbly for a licence. Pilgrimages by IBM bosses were rare. Gary Kildale should have been greatly pleased. He was not. When IBM brought in their lawyers and wanted to prepare a contract Gary got tired of the procedure and kicked them out.

This was Bill Gates' big chance. He had already adapted a computer language to the needs of a small computer so he could confidently promise to deliver that kind of software to IBM, but in fact he did a lot more than that. He promised to deliver an operating system as well. His only difficulty was that he did not have one. Luckily (the story of personal computers seems to be laced with lucky incidents), there was Tim Patterson down the road who had, more or less ready, an operating system not much different from Kildale's CPM. He called it QDOS, standing for quick and dirty operating system. As with all small companies Patterson was short of cash. Bill Gates bought the operating system for the bargain price of $50,000. He did not ask for a licence. He bought it lock, stock, and barrel. He was allowed to do whatever he wanted with it. His next stroke of genius was to lease it to IBM under the name of MS-DOS and not to sell it.

As expected, IBM's launch of the personal computer in 1981 was a great success. Surely, if IBM lends its name to a new generation of computers they are bound to be good. And they were very good indeed. In the first three years they sold 2 million of them. But there was a snag. The new computer was not IBM's own make. They bought the components in the open market, so nothing prevented other computer manufacturers from doing the same. There was though some difficulty with one of the components which was strictly IBM's own design, connecting the hardware to the software. It was patented too. The question was whether the companies who wanted to produce clones (Compaq being the first among them) could crack the function of that vital chip. They could, and they could reproduce it without violating IBM's patent. So from then on any company could produce an IBM compatible machine. Compatible meant that all the software applications (word processors, games, spreadsheets) written for the IBM machine could be used on the clones too. But what about the operating system? Could they copy that too? No, they couldn't. The copyright was owned by Microsoft. Each of the computer companies producing the clones had to pay a licence fee to Microsoft.

IBM did not, of course, like their dependence on a Microsoft licence. When they decided to bring out a new generation of personal computers they opted for an operating system of their own. They called it OS/2. The story that unfolded afterwards is similar to the one in the Bible, though with some minor variations. Goliath had no intention of crushing little David; after all little David had done some service to him in his hour of need. So Goliath, very generously, enlisted David's help in putting the nuts and bolts into OS/2. David had no objection to providing some help but in his free time he went to the gym to practise his skills in the martial arts, particularly in discharging stones from a sling. When he was ready he stopped all collaboration with IBM on OS/2, put his sling into action and smote Goliath. Goliath's wound was not mortal but the poor giant has never quite recovered from the blow. OS/2 died a quiet death. IBM never had a chance to dominate the industry again.

What did Bill Gates do afterwards? He brought out his own operating system under the name Windows. He incorporated a new feature into it, GUI. He borrowed it from Apple without quite having their blessing. Was it morally defensible to do so? Well, he did make some changes to it so that for a patent lawyer it was not entirely the same, but that was not the main point. The main point was that GUI was not Apple's own invention. If Apple had the cheek to borrow it from Xerox PARC, then why should Bill Gates have scruples in borrowing the idea from Apple?

Apple were not idle either. Their new product, the Macintosh (a brand of apple), or simply Mac, was launched in January 1984. It was by far the best desktop computer on the market but it was a bit expensive—about one thousand dollars more than its competitors. The battle from then on was between the Macs on one side and IBM and its clones on the other. Even the terminology changed. A Mac was a Mac, whereas all the others were called personal computers or PCs. Macs, despite their price disadvantage, sold well in the year after their launch but then they halted. However, not for long. Very soon Apple came out with a new feature bought from another computer firm: the possibility of copying on paper exactly what was seen on the screen. This may sound a trivial exercise nowadays, but it certainly was not at the time. Adding to it the recently invented laser printer, it became possible to produce with the Mac copies on paper which were as good as if they had come straight from a traditional printer. Desktop publishing started to flourish and the Mac with it. Anyone with a creative mind had to have a Mac.

The Mac II, launched in 1987, was again a successful product but it did not have a long run either. In 1990 Microsoft brought out Windows 3.0, which had most of the features of a Mac. Without any doubt Macs were still superior, but were they worth the price difference? The three-sided battle between Apple, IBM, and Microsoft ended with a clear victory for Microsoft. It was not only that IBM had to continue using Windows: the whole underlying culture changed. The balance between software and hardware altered considerably and probably irrevocably, to the benefit of the former. As hardware became cheaper and cheaper, the challenge was to make use of all that computing power to produce something useful for business or entertainment—and that became the exclusive territory of software. Microsoft went from strength to strength.[3] IBM had some difficult years.[4]

What matters for the subject of this book is that small desktop computers became easily available in the 1980s and 1990s. At some stage within this period, the idea of transferring information from one computer to another one would have surely occurred to someone. So computer networks would have appeared in due course. In fact, computer networks made their appearance well before personal computers had risen to fame. The story will be told in the next two sections.

How Do Computers Talk to Each Other?

In what form would we expect two computers to talk to each other? It would be quite conceivable for a computer to send a number of pulses down a telephone line and ring a fellow computer. If the fellow computer is not busy at the time it could take the call and use the well-known techniques of sound synthesis to say 'this is AMXZ239' in a clear and loud voice. Our own computer would also identify itself and add possibly 'how are you getting on?' or something equally traditional. After these preliminaries they could then settle down to discussing business in synthesized speech (Fig. 16.2). This is a possibility, but somehow we know that computers would not act this way. They would, of course, need some conventions for introducing themselves (computer people call them protocols); they would need to have some facility to say, 'I beg your pardon, could you please repeat the last sentence?' (some protocols take care of that too). However, it seems unlikely that speech synthesis and polite phrases would be the best way to proceed. So how do they do it?

Communications between computers are bound to be different from those between humans because the needs are different. A telephone line set up between two human beings is occupied from the beginning to the end of the conversation. If you hang up in the middle you will very likely offend your co-conversationist. Computers are less touchy. They do not mind if the information exchange is interrupted at any time. Whatever they have to say to each other is in the form of digital signals and takes place in small bursts. It is a bit like two people playing chess on the phone. One says 'Knight c5' and then hangs up. The other one might call two minutes later and say 'Queen f2, check'. If establishing the connection takes considerably less time than the interval between the calls, then it is worthwhile hanging up and calling again. Since computers

[3] In the five years from its flotation in 1986 Microsoft's share price rose by a factor of 10.
[4] IBM's losses in 1992 were close to five billion dollars, the biggest annual loss ever in US corporate history.

HELLO,.... SO NICE TO HEAR FROM YOU...

Fig. 16.2 Conversation between computers.

talk to each other in small bursts it makes good sense to cut off the communications after each burst. How efficient would that be? It would certainly be more efficient than occupying the line all the time, but still not very efficient if setting up the call takes much longer (say 10 seconds) than the length of the burst, which may only be a fraction of a second.

The conclusion is that the public switched telephone network is not the ideal way for computers to talk to each other. It would be sensible to start from scratch and set up a separate network. Exchanges would of course still be necessary, but they would take the form of computers endowed with some intelligence.[5] They would be able to store and process information. Processing is certainly a good thing, but is there any advantage in being able to store information? There is, as the next example will show.

Let us look at the case of three computers connected to each other via an exchange (often called a node or a router, denoted by N in Fig. 16.3a). Assume that computer A is busy transferring data to computer B. Computer C would love to talk to computer A but the line is engaged. If this was a traditional telephone network, then computer C would patiently wait until the line was free. However, the exchange being a computer, it could temporarily store the information to be sent from C to A. Thus, once A and B stop talking to each other and the lines A to N and N to B are free, the information stored at N can be transmitted to A and at the same time computer C could contact Computer B via the node at N. This simple example shows that the ability to store data would expedite communications. But we can do even better when the messages happen to be long.[6] It is then a great nuisance that the line is occupied all the time and is unavailable to other computers. Hence, the idea is to divide the messages into packets of (say) 0.2-second duration, and make fuller use of the network by sending the packets via routes according to whichever happens to be available. The network that can do that is called a packet switched network. An example will show the advantages.

[5] In a sense this would be a reversion to the conditions reigning in the early years of telephone networks when all exchanges were intelligent. In those times they could do a lot more than simple switching. They had, for example, the means of storing information and delivering it to the right place at the right time. A hundred years ago a subscriber might have dictated a list to a telephone operator asking her to read it out later to the grocer when he returned to his shop after his afternoon nap. Once the telephone network was automated, the subscriber had to wait until the grocer picked up his phone.

A more extreme example I know of happened in the spring of 1944 in Hungary when all the Jewish population outside Budapest was to be deported to Auschwitz. Some of the manual operators listening to official conversations learned about the deportations and warned Jews to move immediately to Budapest.

[6] Long in this context might mean a few seconds.

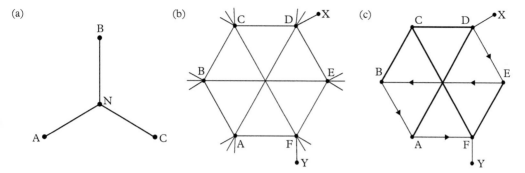

Fig. 16.3 (a) Three computers connected by a node, (b) a partially interconnected network, and (c) arrows showing the routing of a packet from computer X to computer Y. Bold lines are engaged.

Figure 16.3b shows a packet switched network with six exchanges. Each exchange has a number of computers connected to it. The exchanges are partially interconnected. Instead of the possible five, each exchange is connected to only three others. How many ways could one send a packet from computer X to computer Y (usually called hosts) using this network? Since exchanges D and F are not directly connected, the simplest routes would be DEF, DCF, or DAF. There are though some more complicated routes available too, namely DCBAF, DCBEF, DABEF, DABCF, DEBAF, and DEBCF. If the FE, FC, BC, CD, and DA lines all happen to be busy (heavy lines in Fig. 16.3b), the computerized exchanges can still direct the packet from X to Y via DEBAF, as shown by the arrows.

In more complicated networks, the packets may very well travel by a variety of routes and may be held up for shorter or longer times at some of the nodes. So it is quite possible that they will not arrive in the same order as they were sent. So the relevant protocols must take care of that too.[7] They attach headings to each packet which contain information both about the sender and about the addressee, and also indicate the relative position of the packet in the whole message.

It may be seen that even in the simple network of Fig. 16.3b the computers controlling the network have a lot to do. The one at D, for example, has to find out which of the nine lines between the exchanges are available, and whether those open could serve for sending the packet from host computer X to host computer Y. Now imagine that there are not six but tens of thousands of exchanges, and a great many of them need to be interrogated before a decision can be made. Considering further that the information received is sometimes corrupted by noise, it is a miracle that modern communications based on packet switching can actually work. But it can. Noise is, of course, always a problem. Fortunately, modern techniques include a large number of various error-correcting codes which can reduce mistakes below system requirements. An example of a simple error-correcting code was given in Chapter 11.

However, errors due to noise are not the only hazard. There could be errors in the writing of software too. As mentioned previously, a digitalized exchange is nothing other than a set of computers controlled

[7] Protocols have developed gradually. There is one used by the Internet which has hardly changed since the end of the 1970s. It is called the Transmission Control Protocol/Internet Protocol, usually abbreviated as TCP/IP. There is also a body called the International Standards Organization (made up from members of the national standards organizations), which has made recommendations for a framework consisting of a seven-layer protocol. At present there are still a variety of protocols. Harmonization is bound to come sooner or later.

by software. And the program could be faulty. Well, you would say, if there is a danger that something is wrong with the program, then somebody should check it, and if the program is rather complex, then perhaps more people should check it. But how can the program be checked? Apart from simple programming mistakes which can be found easily, one would not know whether the program works as intended or not until it is tried in practice. And even if it is tried in practice, the mistake may not be immediately obvious. The system may work without blemish for years and then suddenly a situation arises with which it cannot cope. One never knows. Has any major disaster happened due to software failure? Yes, there was a memorable failure on 15 January 1990, closing down practically the whole AT&T network.

It started with a computer (let's call it A) having some minor trouble. So it announced to its neighbouring computers, 'please take over my functions for a little while until I recover'. Well, it took a couple of seconds or so for computer A to recover and then it sent another set of signals to the neighbouring computers saying that 'I am in good health, happy to take back my functions.' The neighbouring computers took notice of A's return to service. But due to inadequate software while these computers were busy taking note, they tended to react in an erratic manner to some other incoming signals. Seeing this, the boss computer turned them off. But when, in their turn, they announced their recovery the same mistake occurred again and they disabled some further switches nearby. There was a chain reaction and very soon all the digital switches were out of action. How did the network recover? Not knowing exactly what was wrong with the software, the engineers simply replaced it with an earlier version. They returned to the later version when that program had been debugged.

How complicated has software become nowadays? How many instructions are needed for a digital network to function? My answer in the first edition, published in 1999, was: "miles and miles of them." I introduced there Foster's metric that gives the length in miles, one mile of software corresponding to 400,000 instructions. My prediction for the next ten years was an increase by a factor of ten. Considering that being in 2020, another 10 years have passed bringing the forecast to 300 miles of software corresponding to 120 million instructions. This is about correct. Operating systems nowadays have about that many instructions But the longest software in existence, that for the analysis of the human genome, has over 3,000 billion lines of code-hardly imaginable.

Would a new approach, direct connections to and from the human brain, reduce the number of codes required? It might. In Chapter 17 a few more words will be devoted to this possibility.

A Brief History of Computer Networks

It will come as no surprise that the first collaboration between computers was proposed for military applications. When the Cold War started there was a genuine fear in the US of a possible Russian bomber attack. The proposed action was to set up 23 computer networks with the aim of linking about a hundred radar stations, collating automatically the data about approaching planes, identifying them as friends or foes, and directing fighter planes against the intruders. This system was called SAGE, standing for Semi-Automatic Ground Environment. It was a major and costly enterprise. It cost several billion dollars—more than the Manhattan project. It became operational towards the end of the 1950s but it never worked perfectly (fortunately, there was no need for it). According to some tests at the time it could have shot down only a quarter of the incoming bombers. Nevertheless, it was a useful exercise. It proved that computers could collaborate to achieve a specific aim. It never became operational because new requirements arose when the threat changed from bombers to ballistic missiles.

Civilian operation of computer networks started with flight reservations. In 1959 IBM started to develop such a system for American Airlines. Banks followed soon, first introducing interbank transfers and later automatic teller machines, which did make a big difference to the way we draw money from banks. These were automated systems interconnecting local and central computers but they were employed for a specific task. A much more challenging problem arose when the possibilities of computers working together were considered in a general context. The question was how to share resources between computers, and how a computer can be shared between many users, each one of them wanting it for some disparate purpose.

The whole thing originated with the US Defense Department, who set up the Advanced Research Products Agency (ARPA). The main interest of this agency was information processing with the eventual aim of using the results of the research for military purposes. Perhaps the most important problem they were supposed to address was the design of a network that could withstand a devastating enemy attack. They started with no specifications because nobody knew in the Pentagon, or anywhere else for that matter, what kind of infrastructure they really wanted. So they gave a free hand to a number of creative researchers to set up a computer network and see what all those computers could do with each other.

The man in charge of setting up the network, which rose to fame later as ARPANET, was Robert Kahn of Bolt, Beranek, and Newman Inc. He had to start from scratch because the problem of computer communications was an entirely new one. The project began in the late 1960s. Although there had been suggestions of similar kinds before, this was

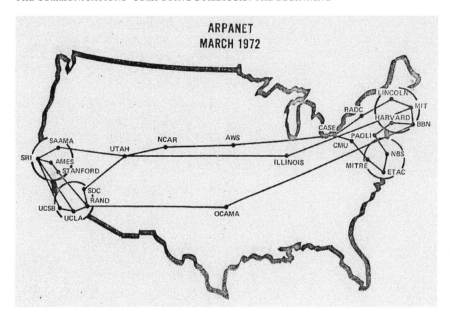

Fig. 16.4 ARPANET in March 1972. Source: Wikimedia Commons.

the first network covering a wide area that was built around the idea of packet switching. It was nearly exclusively a US effort (the nodes in March 1972 may be seen in Fig. 16.4), with two exceptions: there was a node in Norway and another one in London. A link to Norway was actually established well before the advent of ARPANET for the transmission of seismic data to Washington, DC. The aim was to keep an eye (or rather a sensitive ear) on Soviet nuclear tests. The node at London was for research purposes. It was first offered to the National Physical Laboratory on the strength of Donald Davies' already established packet-switched local network. But those were sensitive times politically. Britain had just negotiated her entry into the European Community. New contacts between a British government organization and the American defence establishment were not encouraged.[8] Davies had to decline the offer. His place was taken by Peter Kirstein's group at University College, London, which had for a long time been interested in similar problems. An early result of the collaboration was the use of satellites for conferencing, which meant that each computer site participating in the experiment was connected to all the others.

Peter Kirstein's other claim to fame is that he persuaded the Queen to send a computer message (an email, in fact, although the word had not been coined as yet) and even chose a username for her: HME2. She is shown sending the message in Fig. 16.5. Let us remember that when the transatlantic cable was laid Queen Victoria sent a brief formal message to President Buchanan. The message by Queen Elisabeth II was more specific. It called attention to the burgeoning British software industry. The actual text was

[8] It was remembered that de Gaulle vetoed the UK's entry into the European Community on the basis that MacMillan, the prime minister at the time, negotiated a defence agreement with the US at Nassau.

Fig. 16.5 Queen Elisabeth II sends a computer message on the 26 March 1976. Source: Peter Kirstein.

This message to all ARPANET users announces the availability on ARPANET of the Coral 66 compiler provided by the GEC 4080 computer at the Royal Signals and Radar Establishment, Malvern, England.

Coral 66 was a high-level language adopted by the Ministry of Defence in Britain.

ARPANET was exclusively a Defense Department effort run by civilians. However, not very long afterwards, other nets followed. THEORYNET came in 1977 at the University of Wisconsin, BITNET in 1981 at the City University of New York, EARN (European Academic and Research Network) in 1983, JANET (Joint Academic Network in the UK) in 1984, and NSFNET in 1986. It worked out very well. The Department of Defense got a network to play with and so did the universities.

Progress was similar to that experienced by all successful technological innovations. Slow progress at the beginning, but after a while prices become low enough to encourage the growth of the market, and expanding markets lead to economy of scale in the manufacturing process so that prices decline further, etc., etc. By the early 1980s when personal computers made their appearance every person in research, at least in the developed countries, had already acquired or was in the process of acquiring a computer. Of course these computers were designed to work on their own: nonetheless, the number of those connected to some kind of computer network rose relentlessly. In 1971 the sum total of host computers on the network was no more than 23, increasing to about a hundred in 1975, to four hundred in 1980, and to a thousand in 1984. The biggest boost was probably provided by the establishment of the NSFNET. NSF stands for National Science Foundation. Its main function was to provide resources for academic research in the United States. It was responsible, among others, for setting up supercomputers for the benefit of the research community. Supercomputers are expensive things. So there were only five of them available at the time. The question was how they could be put at the service of any university staff anywhere in the US. An obvious solution would have been to set up a local network at each university and join each one of them to each of the supercomputers. It was an obvious solution but a rather expensive one. Instead, universities were connected to their nearest neighbours, creating a number of networks. The five supercomputers were also connected to each other and then each university network was connected to the supercomputer network. This way it was possible for any university to make contact with any of the supercomputers, although the data might have had to follow rather tortuous routes.

Sooner or later someone was bound to come up with the idea that the various computer networks might somehow be interconnected, and the name Internet easily offered itself. (There were at least five organizations which claimed afterwards to have invented the name.) The main difficulty was that most networks had different sets of operational rules. The problem was similar to that of a hundred different nations, each speaking a different language, wanting to communicate with each other.[9] There are then two feasible solutions: (i) a central authority forces them to use one common language, which may be the language of the dominant country or may be an artificially created language like Esperanto, or (ii) interpreters are employed to translate from each language to every other language. For networks the former solution entails the introduction of a common standard, which will come eventually, but for the time being we are faced with the latter solution in which an interpreter (computer scientists call it a gateway) is provided at the interconnection of any two networks. This is the way the Internet operates.

The US Department of Defense was clearly pleased with a network that connected its various offices to each other. Academics were pleased at the chance of scientific collaboration offered by the networks, and particularly for having access to supercomputers. Could the Internet offer any other service? Yes, it brought us electronic mail. By coincidence, not by design. It could not have happened by design because that would have violated the monopoly of the post, telephone, and telegraph authorities in most countries. However, when two computers were in contact with each other nobody could prevent computer operators from sending messages to each other. As the number of computers on the networks increased, so did the number of people who contacted each other via the computers. After a while it became official.[10] Users could acquire an electronic mailbox to which messages could be sent. The post, telephone, and telegraph authorities surrendered. In fact, the

[9] This was also the reason why that enterprise at Babel to build a super-tower had to be abandoned.

[10] That meant also that countries had to be distinguished from each other by a code. They were allotted two letters for the purpose—except the Americans. They did not need one. They invented the whole thing in the first place. The situation was similar to setting up a postal service by issuing stamps. Ever since every single country has put its name on its stamps—except Britain. There was no need for it. The British were there first.

spread of email accelerated their demise. Authorities which are unable to exercise their monopoly power would have to relinquish it sooner or later.

What was the next great advance? The World Wide Web. It started at CERN, a laboratory situated in Geneva, which employs a large number of nuclear physicists who spend enormous sums of our money on research into nebulous subjects. Their original aim was to discuss their results with colleagues all over the world, sending text and pretty pictures to each other. The man who set up a web for them was a software consultant, Tim Berners-Lee. He wrote his first proposal in 1989 and perfected it by the end of 1990. He adapted a software system called hypertext (conceived by Ted Nelson in the 1960s) to the needs of communications between nuclear physicists. He joined his web to the Internet without provoking much notice. Nuclear physicists were, of course, delighted. Global interaction between nuclear physicists became a practical possibility.[11] But this was again only scientific staff, and so of no interest to the layman.

It was really the next advance which changed the Internet from a research tool into a communications medium for practically everybody in the developed world. Marc Andreessen, a young man at the University of Illinois, conceived and brought into life a browser called Mosaic. It made a tremendous difference to the practical capabilities of the Web. It suddenly became possible to open a new file just by clicking (i.e. by pointing a mouse and pressing a button) at a word or image. That was all one had to do and a new file, irrespective of where it had been residing in the big wide world, duly appeared on the monitor. Nothing could have been friendlier than that.

Mosaic was marketed under the name of Netscape Navigator by the company Netscape Communications, founded by Andreessen. When it went public in 1995 it was valued at $2.7 billion. Surfing the Internet became big business.[12] Microsoft came a little late to the scene. But by an ingenious move, Microsoft incorporated their Internet Explorer into the Windows 95 operating system, offering it for free and starting the 'browser war'. They soon became dominant. Netscape added additional features to its browser to make its product more attractive to buyers, but Microsoft followed suit and appeared to win the confrontation. Netscape appealed to the US judiciary. They claimed that Microsoft used its monopolistic position to put Netscape out of business. Antitrust investigation against Microsoft started in the spring of 1998. A historical trial began in October of the same year. In 2000 Federal Judge Thomas Penfield Jackson ruled that Microsoft had unlawfully maintained its monopoly with Windows and had unlawfully included Internet Explorer into its Windows package. He ruled that the company should be split up. It did not happen. Jackson was taken off the case on some technical issue and his successor, District Judge Colleen

[11] Most of the relevant news became quickly available to the whole community. For example, pictures of collisions from the Large Electron Positron Collider were on the Web just minutes after they had been recorded.

[12] The phrase 'surfing the Internet' is said to have been coined in 1992 by Mark McCahill.

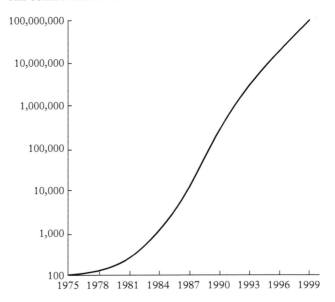

Fig. 16.6 The growth of host computers from 1975 to 1999 on a logarithmic scale.

Kollar-Kotelly, saved Microsoft. A settlement was approved in 2002. The so-called consent decree forced Microsoft to make Windows interoperable with non-Microsoft software and also barred the company from making any arrangement that excluded competitors from new computers. One of the consequences of the anti-trust lawsuit was that Bill Gates resigned as CEO of Microsoft. His place was taken by Steve Ballmer, who was previously Microsoft president. The new boss changed the company culture. Under Gates Microsoft was regarded as a dangerous bully; under Ballmer, it acquired the image of a benevolent giant willing to cooperate.

Let me finish the history of computer networks with a graph showing their initial spectacular growth. As may be seen in Fig. 16.6 the number of host computers grew from about 100 in 1975 to about 30 million in 1998. And afterwards? It grew into a monster. That story will be told in Chapter 17.

The Minitel in France

There is no doubt that, without the efforts of the US Defense Department and of many American academic engineers, the Internet would have never developed. But, in fact, the US was not the only country considering the setting up a network of computers. Well before e-mail became widespread in the US, a deliberate attempt was made by the French telecommunications authorities to propel their country into the electronic age. As discussed in Chapter 11 on digitalization, the great leap forward of French telecommunications started in the early 1970s. Between 1970 and 1983 they managed to increase the number of telephone lines from a desultory 4 million to a very respectable 21 million. A decision was taken in the same period to thrust electronic communications upon the French people. The question was how to persuade potential customers to enter the field. The great idea was to offer an electronic terminal instead of the paper version of the telephone directory—entirely free of charge. That ensured that the number of people taking up the offer was large, which enabled manufacturers to bring them out cheaply. The terminals, fairly simple computers in fact, were marketed under the name of Minitel. An early version may be seen in Fig. 16.7. It provided text only, it was monochromatic, and it had a keyboard. It was connected to a telephone by a modem, and all the information travelled on the telephone lines. It could be downloaded at a speed of 1.2 kb/s. The uploading speed was 75 bit/s. It was designed in the late 1970s, and service to the public started in 1982. In the first year there were 120,000 subscribers, increasing to 600,000 the year after, 3 million by 1987, and 6 million by 1990. It had 25 million users in its heyday. Minitel flourished much before the Internet was available to the public. It worked happily for many years; alas, eventually it could not compete with the Internet. The service was stopped by the French authorities in 2012.

The same kind of technical solution was tried later by a good dozen countries but none of them were successful. The success of the Minitel's operation for many years is somewhat analogous to that of the Telefon Hirmondo (Telephone News, see Chapter 5) launched towards the end of the nineteenth century. Most attempts failed very soon after the launch; only one survived (in Hungary in that case) and even that had to be closed down when telephone news could no longer compete with radio news.

Fig. 16.7 An early Minitel terminal connected to a telephone. Source: Getty Images.

The Impact of Email

Email came in the 1970s. It was not a revolution, rather a slow evolution. It soon reached Oxford. There was no big clamour, at least not in my Laboratory, to use the opportunities offered. There was only one research student who used email extensively and that was because her husband worked in the US and she wanted to stay in close contact with him. Five times a day she used one of the PCs in the lab, to the annoyance of many others who wanted to use the computer for more official purposes. The rest of us in the laboratory viewed it as esoteric behaviour. 'Why don't you write him letters?', she was often asked. 'Because it is too slow', she replied. So why don't you 'phone him?', was the next question. 'Because this is free, stupid!', she answered. This was my first experience with email. It took some time to understand its significance. While we were not allowed to use the telephone for private purposes, email was free and instantaneous, and not even the administrator in the department thought otherwise.

However, is it worth sacrificing the ancient art of letter writing for such a technological upstart? This was a question that had never been asked before because there were no alternatives to the written word.[13] Writing letters was always regarded as the natural state of affairs. All of our ancestors wrote letters (some of those were of such importance that we have been calling them epistles), and there was a time when whole novels were written in the form of letters (see Montesquieu's *Lettres Persanes*).

Although we did not think of it at the time there were always strong arguments against writing letters. First, writing a letter might take quite a lot of effort—and that's not all. Once the letter is written it needs to be put in an envelope. The envelope needs to be addressed and sealed and often reopened because something was left out. It needs to be sealed again only to find that the glue on the envelope has dried out and can no longer be relied on to reseal the envelope. So a piece of sticky tape is required, which tends to stick either to our fingers or to the wrong side of the envelope. Then a stamp of the right denomination must be found. We soon realize that the right amount of postage to Ruritania has just slipped from our mind, and even if we remember, it turns out that no combination of the available stamps would add up to the required amount. Finally, the letter needs to be dropped into a pillar box but when we find the pillar box we haven't got the letter with us, and when we have the letter, there is no pillar box in sight. The use of email is much easier. One types the letter, clicks on the address of the recipient in the address book, and then clicks the send button, and that's all. No envelope, no glue, no sticky tape, no stamps, no pillar boxes.

Apart from being fast and free, email has other advantages too. Maybe the greatest one is the possibility of sending the same message to

[13] This is not quite true, because the fax was an alternative, but it was too complicated to use and, of course, cost money.

a large number of addressees. It is a major advantage for business. They can praise the quality of their ware to a great many people simultaneously. For criminals it is a godsend. They can simultaneously send various scams to a large number of people. It would also be a tremendous advantage for a budding prophet. Instead of visiting city after city, village after village, gathering the faithful and addressing the multitude, he or she could instead send an email to all potential disciples. In science it is a great advantage to be able to have instant communications. Groups in different countries can collaborate as if they were all the time in the same laboratory.

One could also consider as an advantage the psychological pressure upon young men and women who have jobs in other parts of the world. They no longer have an excuse to cut themselves off from the old country, from friends, from Mum and Dad. They are expected, and they mostly comply, to click the mouse and send a few lines now and then. Actually, they can do more nowadays than sending messages. They can send videos of, say, the first hesitant steps of an offspring, or a birthday celebration showing the cake, the candles, and the happy smile of a two year old.

What are the disadvantages? The main one is hacking. A computer is like a castle but instead of high walls it is defended by dedicated software. Unfortunately, castles can be taken and computer security can be breached. One of the worst cases is when hackers get hold of your address book and send emails in your name to all those on the list. There are several versions. One I have seen many times in the past decade is a plea to an addressee to send money to some account because his friend is in trouble. Having been robbed of his wallet he is stranded in a foreign land. A cruder version is just blackmail demanding to send the hacker money (bitcoins); otherwise, he or she will delete all the files on your computer.

Let me just quote one more disadvantage that also has psychological undertones and is related to writing letters. It mirrors my own experience concerning the art of letter-writing. In the past every letter of mine was a masterpiece sprinkled with humour, wit, and acute observations. Nowadays, I tend to stick to the information to be conveyed. I write many more letters but the quality has gone. I presume most email users are in a similar situation. I just cannot imagine that anyone would ever want to publish a collection of emails, unless of course they are highly confidential or serve as instruments of revenge.

And then there is the time factor. If in the past it took four weeks to receive a letter from India, then you were entitled to wait a week at least, possibly a month, before sending a reply. If the post needed one week to get you a letter from Ruritania, then you had a week at your disposal to send your missive. But how long can you wait if nowadays you receive an email from your boss at 9 o'clock at night? You are supposed to reply, or even do some work, before the sun rises. This is the rat race at its worse, and it is email which caused that.

Cybernetics

In this chapter we have looked at the communications–computing symbiosis, revealing the strong links between the two. It is a big subject that we have treated rather briefly. There is, however, another discipline which includes those two, and control as well. We could call it the communications–control–computing symbiosis. It would be a good name although a little lengthy. Other people call it cybernetics. It is unfortunately beyond the scope of this book. Nevertheless, I must mention it briefly, partly because of its intrinsic interest and partly because it played a significant role in the Soviet Union, causing eventually the collapse of the communist system. It is not quite straightforward how the collapse is related to cybernetics, but I shall make the thesis plausible in a following section. In the present section I shall give a very concise description what cybernetics is about.

The word was coined originally in the nineteenth century by Andre-Marie Ampere (the man after whom the unit of electric current is named), who derived it from the Greek word *kybernetes*, meaning governance, steersman, pilot. It was in his essay on civil government, an entity that clearly needs to be governed. Norbert Wiener (born in America of European parents) resurrected the term in his 1948 book, *Control and Communication in the Animal and the Machine*. It is a very general subject. There are many definitions, some of them half a page long but still unable to do justice to the subject. If I had to define it in one sentence and not repeating the description above, that it is a symbiosis between communications, control, and computers, then I would say it is the study of systems and feedback. Systems can have an enormous variety: tt could be a system of computers, a nervous system, a management system, a whole society. Feedback sounds like a technical term, but it is the simplest of them all. Feedback means information feedback. Say your job is to steer a boat. You set the rudder but that is not the end of the job. Some disturbance occurs, say the wind direction is changing. The information about the change of wind direction needs to be fed back. Once it is fed back, you can act upon the information. You swing the rudder to steer the boat back to the original course. Or to give another example: you drive a car. You watch what happens in front of you. Assume that a passer-by suddenly decides to cross the road in front of your car. The new piece of visual information is that there is someone in front of the car. The information triggers a feedback. You need some action. You put your foot on the brake. Or take a thermostat. It senses that the temperature in the room has slightly decreased. The feedback is a command to the boiler to increase water temperature or get the pump going. These are, of course, very simplified examples. If the reader wants a more complicated one, take my favourite: what is the minimum number of components a machine should have in order to be able to reproduce itself?

Science, Technology, and Dictatorships

We have seen in this book that communications is an all-pervading force. It affects society, it affects diplomacy, it affects military matters. What we have not investigated so far is how communications fare in a dictatorship. Well, this is not quite true. We did briefly mention a communications engineer, Guglielmo Marconi, returning to his native country, Italy, during the 1930s, which at the time was ruled by Benito Mussolini, and the famous inventor became a member of the Grand Fascist Council. Mussolini was surely a dictator and his Fascist thugs committed many an atrocity, but it was not an absolute dictatorship. Italians, by nature, are a happy-go-lucky nation, for whom absolute control is absolutely alien.[14]

Germany, in the years between 1933 and 1945, was under a much more rigidly controlled dictatorship. There was no freedom of the press, no freedom of assembly (apart from the Nazi rallies), no freedom of speech; however, the Nazi rule was not absolute. Passports to travel abroad were freely available (not for everybody, of course, but for the large majority). Contacts with émigrés were frowned upon, but many academics managed to correspond with them. It was a murderous regime, but they never reached the perfection achieved by the Soviet Union.

The Soviet situation was best described by the second part of the joke above.[15] In Stalin's Soviet Union anybody could be sent to the gulag at any time. Anybody could be dismissed from his/her job at any time. If the authorities so wanted, everybody had to bear false witness against anybody. How did this affect the sciences? It depended on whether the particular science had any bearing on military applications. Although quantum mechanics was denounced at some time as decadent, cosmopolitan, bourgeois, Jewish pseudo-science, it became respectable as soon as its military significance was realized. With Lavrentiy Beria in charge they produced the atomic bomb in a surprisingly short time. They had the motivation and they had some brilliant theoreticians and technologists. What about other sciences? Biology was an expensive failure. For ideological reasons they believed that acquired properties can be inherited, an idea championed by Trofim Lysenko. Geneticists opposing the idea were, at best, dismissed from their jobs, at worst, executed as enemies of the people. It took nearly 30 years to realize that Lysenko was a charlatan. The damage to Soviet agriculture was

[14] The joke goes, 'In Italy everything is allowed, including the things which are forbidden.'

[15] 'In the Soviet Union everything is forbidden, including the things which are allowed.'

immense. Whether any science was regarded good or bad depended on the views of a few philosophers and ultimately on the philosopher-king Iosif Stalin.

Computers, Cybernetics, the InterNyet, and the Soviet Collapse

We have seen how the Internet emerged in the West. It emerged quite gradually as communications engineers had some original ideas and put them slowly in practice. Military considerations played a very important role. It was the US Department of Defense that set up the first net known as ARPANET. It was the result of cooperation between the military and civilians.

In the Soviet Union the situation was quite different. I saw that with my own eyes. As an engineer, I visited laboratories in the Soviet Union. I was there, maybe ten times in the decade between 1975 and 1985. My visits were driven partly by professional considerations but also by curiosity, wanting to observe the social changes in the Soviet Union. I immediately noticed how few computers they had on the benches where they conducted their experiments. My rough guess was a factor of 10 between the number of computers on display there and those upon the benches of an average research laboratory in the West. Many of those computers were of Western make but some of them were home-made.

So there is no doubt that the Russians produced computers, and once you have computers, the idea that they should be able to talk to one another is no more than common sense. But, of course, the first thing was to have the computers.

What purpose were those computers supposed to have served? Did the Soviet secret service demand more and more computers in order to improve their hold on the population? No, they did not. Why change anything that had been working so beautifully? It is like introducing expensive safety features into a railway system that has not had an accident for decades. And there was another reason for the secret service's opposition to computers. Those rather mysterious devices had facilities for information storage. The last thing the secret service wanted was to enable any Soviet citizen to collect information. True, they could inspect computers in the same way as they had been inspecting Xerox machines ever since they had appeared on the scene, but computers were in a different class. A properly coded computer, aided and abetted by a printer, could appear to any KGB interrogator as poised to print the collected works of V. I. Lenin, whereas as soon as the agent was out of sight it could churn out the text of the latest BBC broadcast or the last seven editions of a samizdat paper. The likelihood is, of course, that they did receive and install computers in their premises but with no more than moderate enthusiasm.

COMPUTERS, CYBERNETICS, THE INTERNYET, AND THE SOVIET COLLAPSE

So who wanted computers? Scientists and engineers. But they did not have much clout. Those who wanted to promote computers were not sufficiently influential. Those who were sufficiently influential had other things on their mind. The Soviet economy had plenty of problems to think about. Who wanted computers anyway? The KGB was luke-warm for it. Did the military clamour for computers? Yes, but even more lukewarmly than the KGB. What about bureaucrats? Did they want computers? Yes, they were for computers in principle but not in prac-tice. Thus, apart from a few pious resolutions, nothing happened. It was not the fault of Soviet computer scientists. They were as good as their Western counterparts. They could design computers; they could not mass-produce them.

The overwhelming Soviet idea was to produce big things: Big ballistic missiles, big satellites into which they could pack as much electronics as they wanted. They missed the revolution of small things. A further dis-advantage was their devotion to Marxist ideology, or rather the power and influence of those who were the practitioners of Marxist philoso-phy. They regarded themselves the high priests of a state religion and had the corresponding mental attitude. They were suspicious of any-thing that came from the West. They not only decried Western prod-ucts; they regarded them as the evil creations of bourgeois science. Cybernetics fell into that category.

Whenever new ideas came from the West, Soviet philosophers were ready to denounce them. That happened to computers too but the ire of philosophers was particularly reserved for any manifestation of cybernetics. Their language was never refined when attacking Western thoughts. Here are some examples taken from Benjamin Peters' book *How Not to Network a Nation: The Uneasy History of the Soviet Internet*, published in 2017. He cites the titles of articles by Soviet 'experts' who were invariably negative, e.g. 'Cybernetics, a Science of Obscurantists', 'Cybernetics, an American Pseudo-Science', and 'The Science of Modern Slaveholders'. Peters also quotes from an article in the Soviet journal, the *Questions of Philosophy:*

It (cybernetics) is a sterile flower of the tree of knowledge arriving as a result of a one-sided and exaggerated blowing up of a particular trait of epistemology.

Similar sentiments were expressed by the authors of the *Concise Dictionary of Philosophy*, published in 1954:

Cybernetics, a reactionary pseudo-science that appeared in the U.S.A. after World War II and also spread through other capitalist countries. Cybernetics clearly reflects one of the basic features of the bourgeois worldview—its inhumanity, striving to transform workers into an extension of the machine, into a tool of production, and an instrument of war. At the same time for cybernetics an imperialist utopia is characteristic—replacing living, thinking man, fighting for his interests, by a machine, both in industry and war. The instigators of a new world war use cybernetics in their dirty, practical affairs.

It was never easy to rebut Soviet theories based on ideology. Trofim Lysenko managed to ruin Soviet genetics and agriculture by his theory of the inheritability of acquired characteristics, supported strongly by Stalin. As mentioned earlier, Lysenko flourished for nearly three decades and was discredited only after Stalin's death. Fortunately, for Soviet science the philosophers' stronghold on cybernetics was weakened by scientists and engineers eager to get involved with that 'obscurantist' science. Two famous (in the West, too) Soviet mathematicians, Alexei Lyapunov and Sergey Sobolev, and a mathematically minded young engineer, Anatoly Kitov, a colonel in the Soviet Army, took up the challenge. They published a paper, the very first paper that dared to address the issue, extolling the virtues of cybernetics and claiming that the theory was particularly applicable to Soviet circumstances. In that they were right. The control and optimization of large systems was exactly what the Soviet authorities needed. Central planning is a key element of any socialist economy where all the means of production are in the hands of the state. It was always a problem to manage such an economy, to prepare plans for 50,000 industrial units and 27,000 farms. Those in charge who devised the plans always reported perfect planning and perfect execution. The consumers who had to queue for the necessities disagreed. The authorities were aware of all the problems; they tried ad hoc methods to improve the situation but those never worked. Now with the advent of cybernetics there was a golden chance to solve the problems.

The Russians' golden chance was noted in the US with alarm in the early 1960s, just as when there was worry over the 'missile gap'.[16] After the Soviet Union's successes in producing an atomic bomb, a hydrogen bomb, and the launch of the first satellite, as well as putting a man in space, some people in the US military genuinely feared that there might be something in Kruschev' boast, 'We shall bury you.' In particular, Arthur M. Schlesinger Jr, one of Kennedy's closest advisers, believed, for a while anyway, that the combination of planned economics with cybernetics was very powerful.

Stalin's successor as the leader of the Soviet Union was Nikita Khruschev, whose previous admiration of Stalin did not prevent him from denouncing Stalin's crimes in January 1956 at the 20th Congress of the Communist Party. Kitov, the man who was the most convinced believer in the power of cybernetics, wrote a letter to Khruschev in January 1959, proposing a transformation from the manual and personal forms of management (the administrative staff at the time stood close to one million) to automated systems based on computers. He proposed a unified automated management system, acknowledging that such a

[16] It was believed that the US trailed behind the Soviet Union in missile technology.

move would lead to drastic reduction in administrative staff and significant increase in efficiency.

More formal propositions came from the not-long-before-established Cybernetics Council. They proposed 'a single uniform system of information and computer service, meeting the demands of all institutions and organisations in the processing of economic information and in the execution of computer work'.

Kitov, Glushkov, and the InterNyet

Anatoly Kitov did not stay idle either (Fig. 16.8a). He had a great idea. It was not an entirely new idea because that was the kind of thing happening in the US; he proposed dual-purpose communication lines accommodating both civilian and military applications. He submitted his proposal to the Kremlin whence it was sent to the military authorities for evaluation. They were abhorred. Whoever heard of allowing civilians to get into the heart of a military establishment? The people most shocked were Kitov's immediate bosses. They were disappointed that a young man of promise, a member of the military establishment, could come up with such proposal. They rejected the proposal out of hand, and just to make sure that they will not receive another such proposal from Kitov, they expelled him from the Communist Party and discharged him from the army. This was a setback, no doubt, but not the end of the cybernetics project. The torch was taken up by Victor Glushkov (Fig. 16.8b), who was by that time the Director of the Institute

Fig. 16.8 (a) Anatoly Kitov and (b) Victor Glushkov, two Soviet computer scientists whose attempts to realise a Soviet internet were frustrated by the authorities. Sources: Computer Museum Moscow and History of Computing in Ukraine.

of Cybernetics in Kiev. In the 1960s he submitted some proposals for setting up computer systems which were lukewarmly received by the relevant ministries. Not much happened. His opportunity came towards the end of the 1960s when the Soviet leadership became aware of the American ARPANET program. They were happy to receive proposals. Glushkov submitted his most ambitious plan, which became known as OGAS, from the Russian acronym for All-Union Governmental Automated System. The main opponents were the ministries in general, and two powerful governmental agencies, the Central Statistical Administration and the State Planning Committee. Neither of them were happy with Glushkov's plan. They were both afraid of losing their monopoly positions of collecting and analysing data, and planning the economy.

Glushkov was much better than Kitov at keeping the bureaucrats on his side. He was not in danger of losing his job. On the contrary, he was showered with medals and decorations. Among others he received the Order of Lenin, the Order of the October Revolution, and the Lenin Prize in Science, and he was elected a member of the Soviet Academy of Sciences at the age of 40. The Institute of Electrical and Electronic Engineers also acknowledged his contribution (posthumously) to the digital automation of computer architecture by granting him the Computer Pioneer Award.

So what happened to OGAS? The plan was watered down to the extent of losing its original aim of creating a single unified management system. It was in nobody's interest to set up a unified system with universal computers. The trend developed in practice, whilst pretending that they proceeded in the spirit of Glushkov's plan, was toward specific solutions using specialized computers. The main culprit was the Ministry of Defense. Every type of weaponry, several hundreds of them, was controlled by specialized computers programmed just for that purpose. Even more importantly there was no attempt to involve civilians. The gap between militaries and civilians remained as wide as ever.

The ministries had their own networks consisting of all the enterprises for which they were responsible. They had good communications within them. But they had no means to contact enterprises which belonged to another ministry. Those computers spoke another language. The ministries were not interested in any kind of unification. However, they could use OGAS for their own aims of improving communications in their own networks. Each of the ministries minded their feudal lands.

Why did Glushkov's proposals for a unique computer network fail? An overall explanation is the presence of a mind-boggling bureaucracy in the Soviet Union that was always ready to quash any attempt at reform. A further obvious obstacle was the lack of any tradition to produce reliable and accurate data. In fact, collecting and publishing data was a dangerous business. The results of the 1937 census found that the

Soviet Union had an unnaturally slow growth of population. These results were never officially published. In fact, they were suppressed and many in the Central Statistics Department were arrested for 'crude violations of the principles of statistics'.

There were, of course, other reasons for the failure:

- The plan formulated mostly by mathematicians was too ambitious and too general, well beyond the American programme.

- Opposition came from many members high up in the Communist Party who were afraid of losing control of the economy.

- Opposition came from ministries who were threatened by the prospect of losing their own networks.

- People in leading positions were afraid of losing their jobs.

- There was no tradition of having free access to information.

- There was no conversation between consumers and producers.

- The telephone lines were of poor quality.

- Innovations were not properly rewarded; the inventor might even suffer if he/she pushed his/her invention too strongly.

The Collapse

There were many predictions, starting in the 1920s, about the demise of the Soviet Union. There were also many reasons why the Soviet Union should have collapsed. What was the main reason for the eventual collapse? Politicians and Kremlinologists, they all missed the signs. They missed them because their thinking was determined by decades of analysis of the Soviet political system. The reason was an entirely new factor: the growth of computers and of communications that I called the communications–computer symbiosis. What we have seen so far in this chapter was the Soviet backwardness in computer technology and their inability to create a unified computer network. This question was examined in an article written in the spring of 1984 by O. L. Smaryl (an anagram of L. Solymar). It was titled 'New Technology and the Soviet Predicament' and published in the East–West periodical *Survey*. Its main point was that the Soviet inability to embrace new technology would lead to the collapse of the system. Why? The series of inefficiencies mentioned so far were mainly in the civilian sector. The inefficiency in the military sector was attributed to the lack of unification in their computer system. The other major mismanagement was that they were 'thinking big' instead of 'thinking small'. They failed to master microelectronics. Moore's Law was not applicable to Soviet technology.

And then came President Reagan's Strategic Defence Initiative, known better as Star Wars. It stipulated that strategically placed American space vehicles provided with laser weaponry would be able to annihilate all attacking Soviet weapons. The Soviet military realized that they had no chance of winning the race. They advised caution. 'If we can't win the war, let's have peace', they advised. And this is what happened. The Soviet Union was willing to sign agreements like the Intermediate Range Nuclear Forces Treaty, which reduced tension between the superpowers. Also, hoping to reverse the downward trend in economy, General Secretary Mikhail Gorbachev introduced *perestroika*, a liberalization of the rigid Soviet system. The other major initiative was *glasnost* (transparency), which intended to show that the Soviet Union has undergone radical change and was ready to renounce the crimes committed in Stalin's time. Gorbachev also realized that the Soviet Army would no longer be able to continue their rule of the East European satellite states. He made the brave move of withdrawing the Soviet Army.

Having failed to master new technology the Soviet Union fell further behind in the arms race. The collapse came suddenly as the top members of the Communist elite (including the prime minister, the Home Secretary, the Defence Secretary, the head of the secret service, and even Gorbachev's deputy) tried to mount a coup against Gorbachev. The coup was defeated but the Soviet Union collapsed and Gorbachev's job with it.

Admittedly, the Soviet Union was unlucky. They could not cope with microelectronics, partly because of its political implications. Had science developed in some other direction, had it turned out to be possible to produce an army of supermen, the Soviet Union would have had a dozen divisions in full armour, while in the West the debate would have still been raging about the ethical questions creating supermen.

Part IV
Communications Galore

<table>
<tr><td>17</td><td></td></tr>
</table>

Satellites Again

Introduction

Satellites have appeared several times in our story. They were introduced in the chapter on microwaves and we returned to them when our main concern was global communications and wanted to explain how a message sent by a phone (mobile or fixed) from some part of the world can travel through the skies to reach an addressee in another part of the world. Satellites have been mentioned mostly in some specific context when they have been employed for some useful purpose. In the present chapter we shall look at satellites in a more general way, beginning with a little statistics.

How many satellites have been put in orbit ever since they became both useful and fashionable? Not far from ten thousand. The number of satellites launched each year up to 2019 is shown in Fig. 17.1. The next question of interest is that how many satellites cruise around the Earth, now in March 2020? The answer is a couple of thousand. Which country has launched the most satellites? Quite obviously, it is the United States of America, followed by China and Russia. How many countries have been capable of putting satellites into orbit? Not many. For that they would need powerful rockets. Which countries would have that? First coming into mind are again the US, China, and Russia but some other countries can also do it like Kazakhstan, which inherited its rocket launching site from the Soviet Union. A somewhat easier question: how many countries own satellites? The only commodity needed for owning satellites is money. It is easy to find a country that, for a not-negligible-sum, would be happy to put your satellite into orbit. The answer to all these questions can be obtained by looking at Fig 17.2. The number of countries owning satellites is unexpectedly high. There are 11 countries which have more than 14 satellites and a further 42 which have less than 14. To know more about satellites we shall now look at some of their properties.

Orbital Periods

The relationship between the altitude of the orbit and the orbital period is given by the equations in Appendix 2. Mathematics tells us that the closer the orbit is to the Earth, the shorter the period is. For example, in a low Earth orbit at 781 km altitude the period is just 100 minutes. In a superstationary orbit, above the geostationary orbit, the period is longer than a day.

Getting the Message: A History of Communications. Second Edition. Laszlo Solymar, Oxford University Press (2021).
© Laszlo Solymar. DOI: 10.1093/oso/9780198863007.003.0017

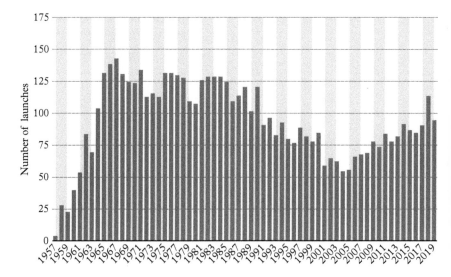

Fig. 17.1 Number of satellites launched in each year. Source: Statista.

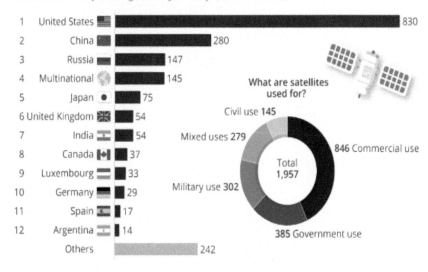

Fig 17.2 Satellites in orbit by country, November 2018. Source: Umion of Concerned Scientists.

The Birth and Death of Satellites

It is well known how satellites are born. They are put on the top of a rocket and moved into orbit at the right height at the right velocity. But how do they die? There are two obvious ways: burn up in the atmosphere or crash into the Earth's surface. These may be accidental or planned deaths. A planned death is agreed at the design stage (as for humans in Huxley's *Brave New World*). Alas, satellites do not live as long as humans, not even as long as cats. An early death is unavoidable because space is such an inhospitable medium. There is a lot of radiation and secondly there is a lot of debris which may again be natural (e.g. meteorites) or may be man-made when bits have fallen off of satellites.

In contrast to the telephone exchanges of the nineteenth and twentieth centuries when young ladies were busy connecting callers, satellites do the same thing but without human intervention. However, that's not the full story. The major damage to satellite communications caused by collisions is the debris produced which might be there for a very long time. Now from birth and death to life expectancy under normal conditions, how long do satellites live? Ten years on average would probably be an overestimate.

Dignified Funeral?

Burning out in the atmosphere can be regarded as the equivalent of a Viking funeral at sea. Half burning out and then falling into the ocean is an even better equivalent. There are, however, some unique satellite burials bearing no equivalence with human funerals. This takes place in orbits reserved specifically for that purpose, called appropriately graveyard orbits. For geostationary satellites such orbits are a few hundred kilometres above their operational orbits where they can rest in peace. The cost of the funeral is a certain amount of fuel needed for the manoeuvre.

I have to mention here a particularly violent type of death that does not occur often. In fact only one case is known where two satellites collided. It happened in February 2009 between one of the Iridium satellites and a defunct Russian satellite, the Kosmos variety. Their relative velocity was 35,000 km/hr. One can imagine a collision of two vehicles at a relative velocity of 100 km/s. That can cause a lot of damage. But a collision at 35,000 km/hr is unimaginable. Presumably, debris can be found thousands of kilometres away. The only set of people who could say something on the matter are particle physicists who investigate collisions of streams of particles moving in opposite directions, each stream with a speed close to the velocity of light.

What happens after a funeral? Life goes on. Those who die must be replaced. How is it done? One possibility is to keep a spare on the ground. The disadvantage is that the process of replacement of a defunct satellite might be lengthy because it still needs to be put in orbit. Another solution is to keep the spare satellite close to its eventual operating orbit. The orbits reserved for this purpose are called storage orbits usually at an altitude of 666 km.

Let us now briefly return to the Iridium satellite constellation. As mentioned in Chapter 9 on microwaves, Iridium SSC faced bankruptcy, but it was then resurrected and is now the most important provider of voice and data communications from any part of the Earth to any other part, including the poles. They still use microwaves (L band between 1616 and 1626 MHz) for communicating both between satellites and between satellites and Earth stations.

They have recently spent a large sum, several billion dollars, on completely replacing their fleet. In the past two years 81 of their satellites were put in orbit by SpaceX from the Vandenberg Air Force base. All of the 66 satellites in 11 orbits have been replaced by new ones, having been changed one at a time. There are 9 further waiting in stand-by orbits, and 6 on the ground. That makes Iridium the most sophisticated satellite constellation which can deliver messages from any point on Earth to any other point, including both poles. Their charging policy also changed. They have now a continuous flow of cash into their pockets.

Satellites that die can be replaced. Humans cannot. The greatest tragedy of space exploration occurred on 21 February 1967. It was supposed to be the first crewed mission of the Apollo 1 programme, the first attempt to put a man on the Moon. The satellite was not even launched. It was during a trial launch that a fire broke out, killing all three astronauts, Gus Grissom. Ed White, and Roger B. Chaffee. There was some uncertainty at the time whether the programme should go ahead but, as we know the decision was to carry on, and indeed 29 months later Neil Armstrong, Michael Collins, and Buzz Aldrin arrived on the Moon.

What's in a Name?

Our planetary system used to have nine satellites, Mercury, Venus, Earth, Mars, Jupiter, Saturn, Uranus, Neptune, and Pluto. This has recently been reduced to eight because Pluto has undergone the ignominy of being demoted to the status of a dwarf planet. With the exception of the Earth they were all named after gods. Therefore, by extension, we might have expected artificial satellites to be also named after gods. This would have been logical and feasible at the beginning but of course with 2000 satellites in orbits we would have rapidly run out of gods. Nonetheless it came as a surprise to me that gods were so much underrepresented. I found only Themis (god of justice) among the American satellites, and Odin (the chief Scandinavian god) as the name of a satellite on which four nations, Sweden, France, Canada, and Finland, collaborated. In addition, the Italians have had Lares (a guardian god) and Helios (the sun god) and the French had Venus (the god of love). I believe that these are the only gods in satellite country at the moment but who knows, fashions can change and gods might re-occupy their rightful places in the skies some time in the future. With those above, all the gods have been exhausted. There have been, however, a few more attempts to avoid the ordinary: one of the first American satellites was called Early Bird. Then there was a French satellite named after Auguste Pickard, who explored the upper atmosphere about a century ago, another one was named after Corot, a renown French painter.[1] The Americans had Kepler, a scientist who, besides doing astrology, discovered some celestial rules in his spare time. They also had Ariel, a little known Archangel. The Danes had another famous scientist, Oersted, and the British had Prospero from *The Tempest*.

One more imaginative name given by the French to a satellite was Asterix, the name of a fat Gaul warrior who with his friends Obelix, Dogmatix, Getafix, and Tragicomix were thorns in the side of the Roman Empire at the time of Julius Caesar. I would though give the palm to the Indonesian satellite Palapa, named after a mythical fruit that the ancient king Gadjah Mada pledged not to eat until all of Indonesia was united.

Unfortunately, most of those in charge of finding names have been hopelessly inadequate in their jobs. As a rule, they have chosen utterly boring names, like the Russian Express-1 which went up to Express-18. Somewhat better were the 31 American satellites all named STARLINK. At least there were references to stars and links. On the other hand, having 31 of them bearing the same name does not say much for originality.

[1] In fact, it is an acronym standing for CoRoT, Convection Rotation et Transits planetaire.

CHAPTER EIGHTEEN

Optical Fibres Revisited

Introduction

Optical communications, including its origins, the major step forward with the invention of the laser, the first thoughts about fibres, and the victory of fibres when both large bandwidth and long distance communications were desired, was discussed in Chapter 12. In the present chapter we shall review the state of the art in 2020, but before doing so let me show a diagram (Fig. 18.1) displaying the progress in the BL product (B = bandwidth, L = length of line) as a function of time.[1] Relative to the electric telegraph lines there has been progress by a factor of about 10^{18}. The relationship started linearly but then there was a kink in the curve when optical lines were adopted. Progress became then much faster. This spectacular progress was made possible with the introduction of new modulation techniques, Wavelength division multiplex (WDM) and space division multiplex (SDM). If we disregard the length of the line and focus only on bandwidth the progress is shown in Fig. 18.2.[2] It can be seen that time division multiplex (TDM) on its own could only account for a more gradual growth.

Following more closely the history of optical communications we should remember that the first lines installed in the 1970s were mostly single-mode fibres. Multimode fibres were only used for very short distance communications. Note that each mode in a fibre has a unique electric field profile and each one propagates with a different velocity. Therefore, at the output of a multimode fibre the fields get mixed up, thereby corrupting the information. Hence, for decades, only single-mode fibres were considered for communications. However, it was later realized that having a number of modes is not necessarily a disadvantage. If it is possible to launch each mode separately and detect them separately at the output end, then this is a way to increase the capacity of the line. Having three modes means the capacity is increased by a factor of 3. Another way of increasing the capacity of an optical cable is by wavelength division multiplex when the same fibre can carry hundreds of frequency bands. Yet more can be gained by space division multiplex when the same cable can carry hundreds of cores.

What other characteristic of optical communication is of interest? We know that practically all cities in the world can be reached by optical lines but it would still be interesting to know the total length of fibres,

---

[1] G. P. Agrawal (2016), 'Optical Communication: Its History and Recent Progress', in: M. Al-Amri, M. El-Gomati, and M. Zubairy (eds), *Optics in Our Time*. Springer, Cham, Switzerland.
[2] Ibid.

Getting the Message: A History of Communications. Second Edition. Laszlo Solymar, Oxford University Press (2021).
© Laszlo Solymar. DOI: 10.1093/oso/9780198863007.003.0018

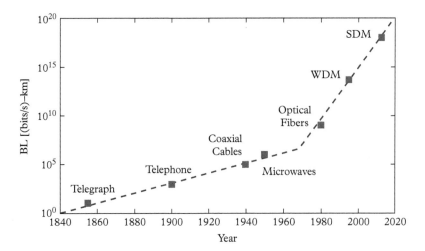

Fig. 18.1 Increase in the *BL* product as a function of time. Source: G. P. Agrawal.

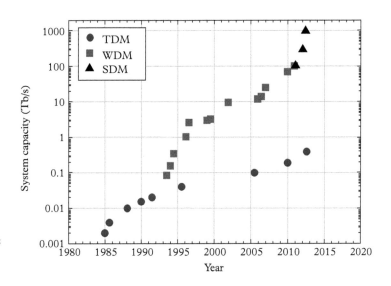

Fig. 18.2 System capacity as a function of time. Source: G. P. Agrawal.

globally. We asked similar questions concerning the transistor count: what is the number of transistors in a chip, and what is the number of transistors produced globally in a year? An answer to the first question was possible, but not to the second question. Nobody knows the total number. It is not measurable. It cannot even be estimated. The situation is roughly similar for optical fibres. While private companies usually tell us the extent of fibres they have installed, governments and particularly the military, keep silent about it. There are, however, further complications: It is no longer true that

One cable = one fibre = one single-mode line.

So are we supposed to count the number of cables, or somehow include the number of cores and the number of modes which may not be known? And what about the 'dark' cores which have been installed but not 'lit'? Would they come into the count? I tried hard to find someone willing to quote a figure but did not find anybody. If I was asked by a friend who knows nothing about the technical details of communications and would expect me to give an answer, I would quickly say: half a billion kilometres.

Fibre Broadband

Considering that the first paper on the feasibility of fibre communications was written by two British engineers, the first town in Europe provided with an optical fibre network was Hastings, and the first experimental fibre optics line in Europe was built in Britain between Stevenage and Hitchin, it is rather sad that Britain has made only limited progress in introducing optical fibres into service. There are lots of misleading advertisements mentioning the word 'fibre' or 'optical fibre', but they only mean that somewhere not too far from your home there is a fibre terminal. The problem of the 'last mile' is still a burning one in Britain.

[3]I cannot escape the feeling that their way of counting fibre connections is not the same as that of the rest of Europe.

The extent of 'fibre penetration' in Europe, the percentage of broadband connections to the home or to the building one lives in, is shown in Fig.18.3. The UK is at the bottom with a mere 1 per cent. Interestingly, the Baltic states are high up on the list and so are Belarus and Russia.[3]

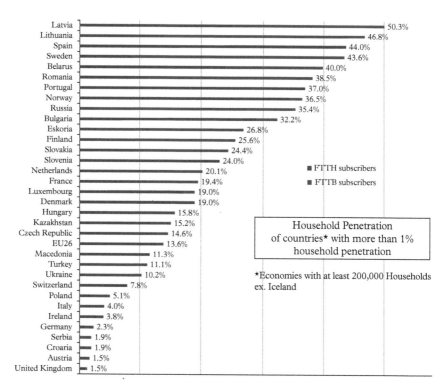

Fig. 18.3 Fibre to the home/building penetration by countries in Europe. Redrawn. Source: FTTH Council EUROPE, March 2019.

Two Major Achievements

We should also mention two record achievements in the field: one by the Australian company Telstra and the other one by Nippon Telegraph and Telephone (NTT). I am well aware of the fact that records in this field are ephemeral. Give a year or two, and the likelihood is that they are all surpassed. Nonetheless they give some indication what the industry is capable of.

Let us start with the one which all communications engineers regard as the first consideration. For Chappe's mechanical telegraph or for microwave links the main design parameter was how far to place the repeater stations from each other. Amazingly, with the invention of the fibre amplifier and the introduction of multiplexing, relay-less long distance communications is possible. How long is long? Telstra Corporation in Australia managed to have an optical line cover the full distance from Perth to Melbourne, over 10,000 km, without a single repeater.

The second milestone I want to mention is a rate of transmission of 1 petabit/s by NTT (the highest point in Fig. 18.2).[4] It comes about as having in a single cable 32 cores, each one having 46 wavelength channels which have bandwidths 680 Gb/s.

It should also be noted that making the fibre out of glass is not necessary. Some lower quality fibres made of plastics are also used for short haul in multimode and for long haul in single-mode operation. Finally, I want to mention 'holey' fibres invented by Philip Russell. They have not been a commercial success but they have scientific interest. Their cross section is shown in Fig. 18.4. The core is solid glass that is surrounded by a large number of holes. The principle of its operation is still the same in the sense that total internal reflection is responsible for the guiding of light but the difference is that the confinement is due to the periodic structure of holes.

[4] 1 petabit = 10^{15} bit.

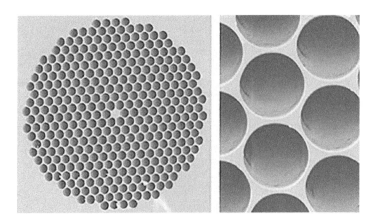

Fig. 18.4 Cross section of a 'holey' fibre. Source: G. P. Agrawal.

Fibres versus Satellite Communications

There is no fundamental reason why fibres could not be used for the same purpose as satellites, that is, communicate from one point of the globe to another point, say by telephone from landline to landline. Assume that the caller is in Europe and the addressee is in the US. Then the connection can be established by the following set of connections: From caller by landline to a European fibre terminal, then by fibre to a US terminal, and finally from the terminal to the addressee by landline. Basically the same set of connections as for satellites. So how to choose one over the other? Well, if the message requires a wide bandwidth, choose the fibre. If delay bothers you, then again choose fibre. In both cases the signal travels a long distance, the speed of propagation in both cases is equal to the velocity of light, but the distance across the Atlantic is much less than that from ground to satellite and back, even for low Earth orbits. Actually, choosing the path the call should take is not up to the caller. It is decided by the caller's broadband provider, and if we think of packet switching, nobody knows how the individual packages find their way to the addressee.

Now looking at the problem through the eyes of the owners of satellites or fibre lines, what they care about is profits. They would realize that with owning satellites the risks of something going wrong are higher, the recurrent costs are higher, and the lifetime is lower than all that for a fibre line, and particularly so if the fibre line can manage without amplifiers. So if you have a choice, choose fibres.

19

CHAPTER
NINETEEN

The Mature Internet

Introduction

At the time of writing (April 2020) the Internet is omnipotent, all-powerful, and all-inclusive. There is nothing that you could not find if you are willing to do a bit of search, and you can get even more if your search turns into proper research. One might say about the Internet that 'when It is good, it is very, very good, but when it is bad, it is horrid.' In other words, there are good sides and horrid sides. I shall start with the good side and will enumerate the various uses of the Internet. I shall enumerate many of the advantages, sprinkled here and there with my personal experience. The logical next step is to look at the horrid side, at the abuses of the Internet. Unfortunately, the abuses are widespread and are getting more and more sophisticated. The list of abuses, as one may expect, is longer than the list of uses, and by the time this book sees the limelight, the list will likely be even longer. How to defend ourselves from the abuses? How can we make the Internet secure? That is a subject upon which treatises and PhD theses have been written but we still do not know who will win, the good guys or the bad guys? Let's hope for the best.

One of the security measures is encryption but that is a double-edged weapon too. It can be used to preserve privacy but just as well to discuss terrorist techniques. I shall briefly discuss some of the methods. I shall also devote some space to statistics about the state and growth of the Internet. Statistics are usually regarded as neutral.[1] They will help to understand the growth of the Internet.

What else should be included? The attempts to censor the Internet are, of course, widespread. The noble aims of ensuring the freedom of the Internet, as expounded by Hilary Clinton, are unlikely to be ever achieved. Not many of the 192 countries in the world are interested in any kind of freedom, let alone freedom of the Internet. In fact, I shall choose only one country, Russia, for two reasons: partly because it is a successor state of the Soviet Union whose approach to human rights left a lot to be desired, and secondly because we have already reviewed the failed attempt at a Soviet Internet, so it is of natural interest to discuss how they are coping now.

Uses of the Internet

We know how the Internet works, well roughly; we talked about nodes, packet switching, operating systems, protocols, etc. in Chapter 16. In the

[1] However, not by everybody. The saying 'Lies, damned lies and statistics' has been attributed to various nineteenth-century personalities.

Getting the Message: A History of Communications. Second Edition. Laszlo Solymar, Oxford University Press (2021).
© Laszlo Solymar. DOI: 10.1093/oso/9780198863007.003.0019

present one we shall leave technology behind and concentrate on the various uses of the Internet but also on the various negative effects the Internet is responsible for. On balance, is it good or bad? Judgment depends on age. Between 3 and 50, everybody is for it. Over 50, people complain about the failings of the Internet, but after some thought they grudgingly acknowledge that it is not all bad.

Where to start? Let me go back twenty-odd years to when I first made a search on the Internet. I was interested in beacons for communications and came across a series of mountain-top watchtowers used by the victors at the end of the Trojan War to announce victory.

Let me give one example of how I managed to find an item on the Internet using search engines. In Chapter 2 I mentioned a play written by Aeschylus in the fifth century BC. It began with a watchman's soliloquy during which a beacon lit up. This signalled the early arrival of Clytemnestra's husband from all that fighting at Troy. Since this soliloquy was very relevant to the purpose of this book I planned at the time, I wanted to have a look at other translations as well, and decided to use the Internet. First, I used Microsoft's MSN. I typed in the title of the play, 'Agamemnon'. It gave me 4,827 websites. I looked at a few. Some were about pop-music, some about pornography. I did not stop to find out what their relation to Agamemnon was. I refined the search by typing in 'Agamemnon and Aeschylus'. That still left me with 531 websites. Still too many. So I changed to another search engine, Lycos. It yielded no replies whatsoever. Maybe they had a particular dislike of Agamemnon. So I turned to Yahoo. Typing in 'Agamemnon' yielded a much more manageable 9 entries, one of which turned out to be the complete text of the play translated by D. W. Myatt. I then returned to MSN and typed in 'Agamemnon and Aeschylus and Myatt'. Three replies appeared on my monitor, one of which gave not only three different translations of the play I was looking for, but also translations of all the extant plays of Aeschylus with plenty of reference articles attached. The whole thing took about 20 minutes. I presume it was beginner's luck. Other times I was not so lucky.

After that success I reconciled myself with the Internet. I became less reluctant to use it but still did not think of wider implications. I did not ask the question: 'What is the Internet for?' The first time I took due notice was when Hilary Clinton, Secretary of State in the Obama administration, made a forceful point about the Internet. She said,

Both the American people, and nations that censor the Internet should understand that our government is committed to helping promote Internet freedom.

It is a noble aim although I doubt its ultimate success. There will always be governments that will try to protect their citizens from the evil influence of Western imperialism. And the number of these governments will probably grow in the near future. The present president of the

United States quite likes dictators. He is not the one who will lead such a crusade. What else from Hilary Clinton? She liked the open availability of the Internet. She regarded it as a great democratizing asset. 'Once you are on the Internet,' she said, ' you don't need to be a tycoon or a rock star to have a huge impact on society.'

There is no doubt that she was right. The Internet is a democratic institution. Anyone can upload anything legal on the Internet, but not everyone thinks it's a good thing.[2] Umberto Eco, a philosopher and the author of the medieval thriller *The Name of the Rose*, was not in favour. In his opinion, 'Social Media gives legions of idiots the right to speak when they once spoke only in a pub after a glass of wine.... It's the invasion of the idiots.' This isn't much new. History is full of people who were not willing to suffer fools. Some political scientists even believed that the vote of the stupid and the dumb should count less than those of intelligent people.[3] But these theories have been discredited. There is a consensus now that all votes should be of equal weight and the Internet is for everyone.

Top Ten Uses

Let me try to draw up a list. It will not be a list based on statistics rather on my subjective opinions and personal experience, Obviously, the most important use of the Internet is email. Every day, as many as 300 billion emails are sent from one point to another one, all over the world. Its virtues were extolled in Chapter 16. In any list I would expect email to appear as the number 1 benefit of the Internet.

What would be number 2? There are many candidates for that position. I have heard that *booking tickets* is often mooted as the second most important benefit of the Internet. I think this is too much of an utilitarian approach favoured by an influential minority of jet-setters. Many people would dissent, particularly those miserable ones who struggle when they have to do any booking on the Internet. They yearn for the good old times when all that was done by travel agents. I feel for them and would not put booking tickets into a high position.

What would be my choice? Where do my sympathies lie? I was an academic all my life. I appreciate research. But I can see that to put academic research into second position would only show my prejudice. I opt for searching but searching in a casual manner. My main criterion is having fun, having innocent fun, void of any utilitarian aim. My number 2 is *search the Internet for interesting but useless knowledge*. This is, by definition, just curiosity-driven research. For example, the question may arise while you chat with your friends about racism, 'Who was the black athlete who won four gold medals at the Berlin Olympics?' You look it up. You half-remembered it but when you find it on the Internet (yes, it was Jesse Owens) you are happy. Your curiosity has been satisfied,

[2] I am not sure what is illegal. The only example I know of is child pornography and that is not a controversial issue. What is legal and illegal in dictatorships, that is another question.
[3] John Stuart Mills was one of the proponents of multi-votes.

and your friends will look at you with respect. I have another example. The tune of a pop song from your teenage days goes round and round in your head. You can reconstruct two lines of the lyrics. You reach for your mobile. Click a few buttons and bingo the lyrics are found.

What is number 3? Having talked to many young men and women I would put *Internet dating sites* as the third most important benefit. In the old times, a few decades ago, the large majority would not have considered registering with a marriage bureau for help. It was not done. It would have carried some stigma (Is that the only way he/she can find a partner?). Nowadays it is accepted. Actually, more than accepted. Having large numbers of both genders on various lists, the owners of the websites can do meticulous matching of looks and interests. They have quite high success rates. They spread happiness. The dating game has radically changed for the better.

I would put *games* into number 4 position. I think games are the same kind of things as useless knowledge. They are both for fun. However, games have one major disadvantage. They can become addictive. A child or an adult might spend many hours playing games. Such addiction may be at the expense of social life; it might even cause eating disorders or sleeping difficulties.

My number 5 is *online shopping.* Although I try to keep myself away as much as possible from any kind of shopping, I must acknowledge its benefits. Even for those who are not good at quickly making up their minds, it saves time and usually saves money too. Booking tickets is also a kind of shopping so I would include that in the present category.

I would take *search for useful knowledge* as number 6. Searching the Internet is not quite the same thing as immersing oneself in dusty libraries. Somehow the feeling is different and the smell is not the same. In the library you feel yourself a conquering hero. On the Internet it just seems a repetitive exercise, perhaps because you press the same buttons again and again, instead of taking books off shelves. On the other hand the *Eureka!* moment is the same. Information comes in small steps whatever the method used to look for it. The first few attempts are usually far from the target but one can gradually refine the search and find more and more relevant websites/books. It's useful, there is no doubt about it, and we should be satisfied with that.

For number 7 my choice is *education.* Once upon the time there were private tutors, then came public schools, state schools, and then the Internet. For school-age children the Internet offers help with their studies, the help in which they may or may not be interested. Above 18, however, the potential of the Internet for acquiring knowledge is more and more utilized. One can find plenty of reference books on practically every subject, and if reference books are found to be too obtuse, then one can find instead tutorials or even courses covering the subject in detail.

Now the reader might ask why I put search for useless knowledge into the second place, search for useful knowledge into the sixth place, and education into the seventh place. Aren't all three the same thing? Anyway, how can one distinguish between useless and useful knowledge? To my mind the chances of being able to find quick information, however irrelevant and however useless it is, occupies high ranking. Before the advent of the Internet it was very hard to find such information. Let me return to my previous example: 'Who was the black athlete who won four gold medals at the Berlin Olympics?' You would not have known where to turn for information. You would have eventually found it if you persisted but probably you would not have persisted. Nowadays it is easy to find such a piece of information practically instantaneously. Since this kind of question arises to all of us several times a day, I think to find quick answers adds to human happiness.

Search for useful knowledge is lower down on my list because its value cannot be taken for granted. The old method was much slower but safer. Similar is the case for education on the Internet. Yes, there are wonderful new opportunities but then the old methods produced good results as well.

I do not feel strongly about any other Internet facilities. If I had to choose three more to reach ten, I would give, not in any order, *discussion groups*, *online publishing*, and *online banking*. I am not saying that they do not represent progress relative to the old methods (I have even used online publishing quite extensively), but, again, I could live quite happily without them.

Abuses of the Internet

Anything that exists can be abused, proven abundantly by humankind in the past ten thousand years or so. It applies to the Internet as well. I cannot think of anything else that could be as easily abused as the Internet. One should not blame computers for that. They bear only a slight responsibility. The main hazard for a computer connected only to the mains is either a frustrated user who repeatedly kicks it or an old-fashioned burglar who snatches it from its resting place. However, as soon as a computer is connected to a network, the chances of abuse by various miscreants increases enormously. Motives vary. Politically inspired hacking occurs less often but it is the most dangerous one. When it is done on a large scale it can be very efficient and it could change the world. When it is done on a small scale by just sending hostile missives to political opponents, it degrades the political process and undermines democracy.

I shall start with some major abuses of the Internet, with Edward Snowden, an ex-employee of the CIA, who made no secret of his actions.

He copied and made public thousands of confidential emails. He claimed that he had done everything for the benefit of mankind and for the greater glory of open politics. At the same time he took the precaution of leaving US territory, flying first to Hong Kong and from there to Moscow where he enjoys political asylum.

More often in politically inspired hackings the hackers want to remain anonymous although their identity could often be guessed on the basis of *cui bono*, i.e. who benefits from it. Russia was blamed a number of times for hacking into computer systems of countries with which they had a dispute at the time. A few of these attempts are briefly discussed here.

Russo-Georgian War, 2008. This was the first time that cyberattacks were launched during the pursuit of a conventional war. One of those hackings managed to intrude into the website of Mikheil Saashkavili, the president of Georgia, replacing his image with that of Adolf Hitler. Needless to say, the Russians won the war.

Russian hacking of Estonia's infrastructure, 2007. This is known as the Conflict of the Relocation of the Bronze Soldier. The cause was that a statue commemorating the liberation of Tallin, Estonia's capital, by the Russian Army in 1944, was relocated by the Estonian government to a much less prominent place.[4] The cyberattacks targeted Estonian infrastructure, the Parliament, banks, the media, etc.

Democratic Convention hacking, 2015–16. Twenty-seven thousand emails of the Democratic National Committee of the US were leaked mainly by the organization Wikileaks. The likely aim was to damage Hilary Clinton in the presidential race and help the campaign of Donald Trump, who was the Kremlin's favourite. The leak caused serious embarrassments to the Democrats by revealing that the national committee was not neutral in the selection of the presidential candidate, that many of the leading committee members derided Bernie Sanders' campaign. Sanders received a formal apology and several members of the committee had to resign.

Let me add one more hacking with a political background in which Moscow's hand is not suspected. It was a cyberattack on TV5Mond, a French broadcasting corporation. It was perpetrated by a group with contacts to the Islamic State of Iraq and the Levant, a terrorist organization. They managed to get the station off the air for several hours and continued causing problems for a couple of days.

It would be wrong to assume that the US and her allies are always innocent victims of hostile hackers and refrain from similar moves. We certainly hear less of them. I know only of one major interference with computer systems that was perpetrated by the West. It is the Stuxnet virus, reputedly a joint US–Israeli work aimed at disrupting Iran's nuclear programme.

[4] A government which believed that 'liberation' was not the right word.

Another kind of interference into daily politics is to send nasty emails to your political opponents. This tendency has gathered strength recently with the emergence in Britain of the extreme left in the Labour Party. One of the Labour MPs at the time, Luciana Berger, received several hundred hate messages in the course of three days just because her political views were different. While extremists in politics may lose ground, sending hate mail may decline but will never disappear. As long as it is possible to send hate mail anonymously, the practice will be kept up.[5]

Next we can look at another kind of hacking, something of daily occurrence. It is an integral part of modern communications made possible by technical advances. I don't think the telegraph was a medium favouring trolls but the appearance of telephones certainly gave brilliant opportunities to an army of miscreants to make obscene calls. Telephone also had the advantage of anonymity although it was technically easier to find the culprits. Another set of telephone calls I know of were designed solely to annoy the recipient. In my childhood a favourite scheme in Hungary was to call an innocent subscriber in the name of the telephone company and ask him or her to measure the length of the telephone cable. When the unfortunate victim reported the length of the cable, the caller advised him that it was just the right length for hanging him high.

Many of the modern versions of hacking are also designed for causing nuisance in the better scenarios, and causing distress in the worse ones. The milder kind is mainly done by teenage hackers who use it as a measure of their hacking skill. If they can enter the network of the Pentagon or of the White House, that would earn them unparalleled respect and glory in their peer group. This is quite similar to climbing mountains. It is dangerous, it has no utility, and it is very difficult to do. The satisfaction is in the fact of just doing it, and the acquired reputation is an additional bonus.

It is much more sinister when the aim is to cause distress. I have already given an example with the ordeal of Luciana Berger. The majority of such abuses are political or racial but in today's divided world anything can happen; anybody could be sent death threats for any reason.

I have not mentioned so far a vast array of abuses which aim at the enrichment of the abuser at the expense of the victims. It can take a number of different forms. The following list is in no particular order.

Addiction. In the old times this referred mainly to drugs like heroin. Today, it is much more likely that the cause of addiction is the Internet. The symptoms are neurotic watching of social media blogs, obsessive gaming on the internet,[6] and compulsive buying online.

Reading victim's files. The hacker enters the victim's computer, and searches for information, possibly credit card details and passwords,

[5] Hate mail is against the law but very few perpetrators are ever prosecuted because of the difficulty in identifying them. Several of Luciana Berger's abusers ended up with custodial sentences. One of them was convicted for racially aggravated harassment and sentenced to two years in prison.
[6] There were recent reports of a man gambling away his house.

which are then used for drawing money from teller machines or buying merchandise.

Denial of service. This is essentially disrupting services to a website or network by flooding it with spurious requests. Ransom may be asked for stopping the attack.

Distributed denial of service. This is the same as denial of service but the flooding originates from a large number of sources, making it difficult to stop the attack by blocking the sources.

Altering victim's files. The hacker takes hold of your files, installs a virus, and demands a ransom for releasing them.

Child grooming. Contact vulnerable children by email, gain their trust, arrange a meeting, shower them with gifts, all with the aim of sexually exploiting them at a later time.

Sexting. It is the practice of sending sexually explicit images over the Internet. This might be all right between consenting couples but it is open to abuse. One of the abuses is to resend the images far and wide without the consent of the original sender. This is called revenge porn.

Fake news. This has not arrived with the Internet. There have been plenty of examples in the past hundreds of years before the Internet was even thought of. One of them, since telegraphic communications was involved, I described in Chapter 4 where I discussed the confrontation in 1898 between Kitchener, who wanted to establish a red band on the map of Africa from Cairo to Capetown, and Marchand, his French opponent at Fashoda who wanted a blue swathe from Dakar to Djibuti. Kitchener had access to the telegraph; Marchand did not. Kitchener sent a number of misleading reports to Europe, claiming that Marchand was in a desperate position and wants to return to France. The French government climbed down. Britain acquired the red band from north to south in Africa, thanks to the spreading of fake news.

Modern use of the term came from President Trump, who has often complained that his political opponents spread fake news about him while he himself is a paragon of virtue. There is no doubt that the Internet is the best forum if you want to spread news, fake or true.

Spoofing/phishing/smishing. The attacker disguises himself by sending messages from a false source very similar to one you trust, e.g. from the government or even from your GP, and requests personal information.

Piracy. It used to be a crime committed on the high seas by robbing ships carrying valuable merchandise, preferably gold and silver. While this practice is no longer defunct (the entry to the Red Sea is a particularly dangerous place), it has been overshadowed by a different kind of piracy. It is still robbery but robbery of intellectual property which is somehow regarded less odious. I have myself been at the receiving end. Three of my books are available on the Internet, free to download by those who have the expertise.

Pornography. The availability of explicit sexual images on the Internet which could corrupt young children.

Social exclusion. I shall return to it in Chapter 20. It is more related to mobile phones than to the Internet. Ultimately, it is, of course, the Internet because social media blogs appear there, although nowadays they are accessed by a mobile phone.

Stealing identity. This is a particularly nasty kind of theft. Perpetrators get hold of a person's identity documents and are then able to access all of the victim's assets.

I have not put blackmail into a separate category because most of those above involve the extortion of money by threats.

The good part of the Internet is accepted without giving a further thought to it. It is the bad part that worries us. Cybercrime is now likely to be the most frequent criminal activity and is growing. Countermeasures are expensive. The police haven't got the resources for fighting it.

A breakdown of the cybercrime figures for England and Wales, by the Office for National Statistics for the period of one year to September 2018, is shown in Table 19.1. The figures are likely to be underestimates because many victims are reluctant to report the crimes.

Table 19.1 Online crimes in England and Wales for the year to September 2018.

Type of crime	Number of cases
Harassment and stalking	56,561
Obscene publications	11,928
Child sexual offences	9,543
Blackmail	3,639
Other violence against the person offences	2,248
Public order offences	2,475
Sexual offences (exc. child sexual offences)	676
Criminal damage and arson	307
Other offences	3,640
Total	91,017

6	2	Y	J	T	7
A	Q	L	1	5	U
0	E	W	R	D	8
V	G	3	F	4	I
X	M	B	S	P	0
C	H	N	Z	9	K

Fig. 19.1 Polybius' square.

Encryption

If the aim is to prevent hackers from reading your files whether resting in your computer or being in transit, the logical thing to do is to encrypt your files. Encryption has, of course, been practised for millennia. One example goes back 22 centuries to Polybius, whom we have already mentioned in this book (Chapter 2), quoting his coding of the letters of the alphabet so that any message can be transmitted by a set of torches. The idea is to write the letters in a matrix of, say, 6 × 6 where each square represents a different number or letter (see Fig. 19.1).

There are altogether 26 letters and 10 numbers.[7] Hence each letter or number can be encrypted by two numbers. For example, H is encrypted by 62, meaning it is in the sixth row and the second column. The number 8 is encrypted by 36, meaning that it is in the third row and sixth column. If such a ciphered message is then sent to somebody, the recipient must have had a copy of the code to decipher it. We don't know whether this code was ever used. I hope not because it is easily breakable: every symbol is coded by two numbers, always by the same numbers. Hence it can be decoded by noting the lengths of the words and the known frequency of letters in any text. A little ingenuity might also be needed.

It is easy to devise a code that is simpler than that of Polybius'. Instead of coding with two numbers one may code by a single letter. A variety on this one is the code + N, in which a letter is replaced by the Nth letter above it in the alphabet. Thus, each letter of the alphabet is replaced by another letter. I played this game with my friends when I was a child. Having a text of hundred words it took the winner between half an hour and one hour to break the code. We also experimented with a code a lot more difficult to break. We called it the 'periodic code'. The simplest version has a period of two units. I show an example: take the phrase

JACK STINKS

Choose a period having two units, 3 and 5. These are then the two numbers to be written periodically under the text as shown here:

JACK STINKS

3535 353535

This means that the letter J needs to be replaced by M, a letter 3 places above it in the alphabet, and the letter A has to be replaced by the letter F, 5 places above it in the alphabet. Similarly, C and K must be replaced by letters 3 and 5 places above them, respectively. Then having pursued the same method throughout, the phrase is coded as

[7] We could do some saving by using the binary instead of the decimal system, only 0 and 1.

MFFP VWLSNY

Breaking a code like this was much more difficult. We did not succeed often. Imagine then a period of 5 or 6 units; they would be unbreakable by amateurs.

Code-breaking rose to great importance during the Second World War. It is often said that the war was considerably shortened by the British breaking the German Enigma machine. After the war mathematicians, mainly algebraists, on both sides of the Iron Curtain spent many an hour looking at its intricacies. However, since the middle of the 1970s code-breaking has shifted gradually into the civilian sphere. The great invention was the 'two-key' system. That overcame the problem of having to send a key for decryption. In the new system everyone has got two keys: a public key and a private key. The public key is known, the private key is kept secret.

So how does Joe send a secret message to Mary? For the first encryption Joe uses his own private key, and then he encrypts the whole thing again using Mary's public key. When Mary receives the jumbled-up message she uses the reverse procedure: first, she uses her private key and then Joe's public key. Let us say that Harry intercepts the message from Joe to Mary. He, like everybody else, has the public keys of both Joe and Mary. Couldn't he use Mary's key for the first decryption and Joe's key for the next decryption? The beauty of the new method, based on certain mathematical niceties,[8] is that the key that encrypts the message cannot decrypt it. There is certainly a correlation between the public key and the private key, but their use is not reversible.

Now that it is easy for people to send secret messages to each other, might not the techniques be used by terrorists and other criminals? They most certainly use them. All they need to do is go to the local store and buy the relevant software package. So should encryption be illegal on the Internet? All governments want to ban encryption or, at least, give a chance to the security departments to be able to decipher them. This has not happened as yet in the West but it might happen in the future if the security situation deteriorates. In authoritarian countries encryption is regulated; The last word is always that of the security services.

Anonymity

We have discussed encryption, whether it should be allowed or banned. Anonymity, or publishing under a pseudonym, is in the same category. The author just hides his/her name. There might be several reasons for that. We have mentioned one of them in Chapter 4. A proposal for an electric telegraph was proposed in 1753 under the initials C. M. He feared that if he published under his own name his neighbours will think he was a magician. The French novelist Amantine Dupin used the pseudonym

[8]There are certain mathematical problems which can be easily calculated in one direction but only with great difficulty in the other direction. A simple but illustrative example is the product of prime numbers (a prime number is divisible only by 1 and itself), e.g. 29, 31, 113, or 223. The product of these numbers is easy to calculate with a pocket calculator. It is 22,653,901. Ask, however, the reverse question, namely which are the prime components of the above 8-digit number and it takes much longer to find the answer.

George Sand at a time when a lady novelist was a rarity. Jane Austen also published under a pseudonym but had no intention to hide her gender. Very modestly the author of her books was only known as *a Lady*. However, most of the reasons for anonymity were political. Thomas Paine's *Common Sense* was published anonymously in America in 1776. He feared the wrath of the British government. Having got their independence and their constitution, the Americans quickly passed the First Amendment that ensured the freedom of the press. For the first time in their history it is now (writing in 2020) under threat. I must also mention in this context the case of Daniel and Sinyavsky in the Soviet Union who had the audacity to publish in the West under pseudonyms. They faced trials in 1965 for their characters libelling the Soviet state and received long prison sentences.

Concerning the Internet the same arguments apply. So far the freedom of the Internet has been respected apart from cases of hate mail, extreme violence, child pornography, and terrorism. I hope this state of affairs will continue.

Censorship of the Internet in Russia

Coming back to the joke in Chapter 16, 'in the Soviet Union everything is prohibited including the things which are allowed', one might be permitted to believe that there will be problems in the post-Soviet period. When the Soviet Union collapsed some of the optimists believed that Russia will somehow adopt democracy, there will be a multiparty system, free elections, freedom of speech, freedom of the press, and freedom of assembly. I even met some Russians who believed that give a few years and Russia will declare her intention to join the European Community. The pessimists believed that Russia will gradually adopt most of the restrictive laws of the Soviet Union, that it is only a question of time.

The optimists were, of course, wrong. Russia is predestined by the ghosts of history to be an authoritative country. A pessimist I met thought that his worst fears had not materialized. Elections are fairly free, and opposition newspapers still exist, so he is quite happy. Information is, however, too sensitive a thing to leave uncontrolled. In the beginning only websites related to drugs, suicide, and child pornography were blocked, which could happen in the West as well. But in Russia there has been a change since 2013: an increasing number of websites have been blocked. The aims were obvious. Calling for 'illegal meetings' (no demonstrations against the government), 'inciting hatred' (no criticism of the government), and 'violating established order' (anything aimed against the government) 'weaken cultural values' (don't mess with our interpretation of what is good or bad). And of course there is no such thing in Russia as Internet anonymity. All users are monitored. Those using Internet cafes must provide passports.

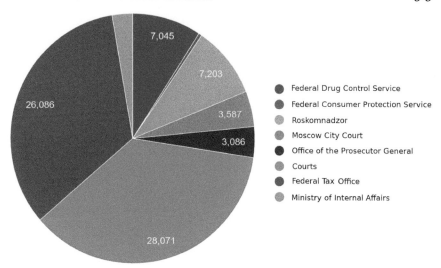

Fig. 19.2 Websites blocked by various Russian authorities. Source: Wikipedia.

There is also a law that anyone showing 'blatant disrespect' online for the state, authorities, the public, the Russian flag, or the constitution can be fined up to 100,000 rubels.

For the number of websites blocked by the various authorities as of June 2017, see Fig. 19.2.

The latest hardening of the rules came in the winter of 2020. It is now law that any part of the global Internet can be shut down in case of an (unspecified) security threat.

20

Mobile Phones, Smartphones

Introduction

We have seen in Chapter 14 how mobile phones came about. The first Motorola mobile phone cost almost four thousand dollars, obviously not designed for mass consumption. It was a commercial venture that was hoped to become profitable some time in the future. Early estimates were low. Some of the optimists believed that with a bit of luck they might sell half a million copies. Ten years after the appearance of the first commercial mobile phone, in 1995, the penetration into the UK was about 7 per cent. By 2004 nearly every adult had a mobile phone. What was the reason? There is a consensus on the matter: improved technology and drastic reduction in price. But this is only the usual process all innovations follow. If you try hard, particularly in electronics, you will always find better and better solutions. If, in addition, you are chased by rivals, that would further improve your performance. But this is still only a partial explanation: a necessary condition but not a sufficient one. Another reason is that a mobile phone is a useful thing to have. Usefulness certainly counts for something, but in today's world it is not even a necessary condition. Let us remember the hula hoop and the Rubik's cube. Neither of them had much utility but they sold like hot cakes, maybe better. Tens of millions were sold.

Let us look at some other possible reasons. A Darwinian one would suggest that there is an atavistic desire to communicate with our fellow human beings. The survival of *Homo sapiens* depended crucially on collaboration. They could punch above their weight.

What else? The emergence of yuppies? They appeared in the late 1980s just about the same time as mobile phones. Mobile phones suddenly became status symbols. The mobile phone went well with the new Jaguar.

Pressure from below? By 'from below' I mean children. Any respectable 10 year old would feel deprived if he/she did not have a mobile phone. I feel sorry for the parents. All resistance is in vain. The only solution is abject surrender.

Fashion? Everyone has a mobile phone. I must have one too. A powerful argument.

Whatever the reasons are, the results have been amazing. It depends a little on how we measure it but roughly, there are as many mobile phones in the world as people. No device, useful or otherwise, has ever reached such distinction.

Getting the Message: A History of Communications. Second Edition. Laszlo Solymar, Oxford University Press (2021).
© Laszlo Solymar. DOI: 10.1093/oso/9780198863007.003.0020

A Few Statistics

The first telephone lines were installed in Boston in 1877 (Chapter 5) and spread quickly around the world. They flourished for a good century. And then came mobile phones: a modest stream first, turning into a flood. The relative number of fixed and mobile subscriptions for the period 2002–18 are shown in Figs. 20.1 and 20.2 for the UK and for the World, respectively. It may be seen that landlines did not give up easily. It was only around 2001 that mobile phones overtook fixed phones. Nonetheless, the writing is on the wall. Whatever their merits landlines will perish.

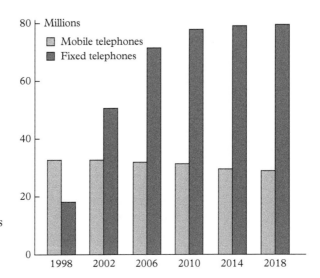

Fig. 20.1 Development of fixed and mobile subscriptions in the UK for the period 1998 to 2018. Redrawn. Source: World Bank.

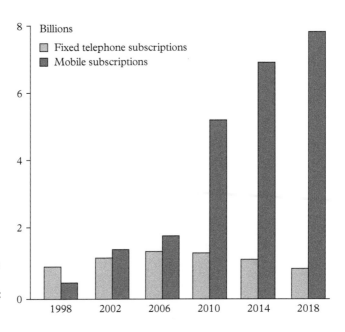

Fig. 20.2 Development of fixed and mobile subscriptions in the World for the period 1998 to 2018. Redrawn. Source: World Bank.

The first mobile phones were just telephones. They could do no more than their immobile counterparts. The first major advance showing their versatility was the addition of a camera (Japan, 2000). From then on they became smarter and smarter until they became known as smartphones. So the next race was between mobile phones and smartphones. A certain fraction of the population, those no longer in their early youth, clung to their mobile phones, happy that they could send and receive a call wherever they were. However, the younger generation wanted more. They, or their parents, bought them smartphones. How they fared is shown in Fig. 20.3 where the number of smartphone subscriptions is shown for the years 2016 to 2020.

Now we can ask the question about the different 'mobile state' of various countries. This is shown for 18 developed and 9 developing countries in Fig. 20.4. Note that there are three items in this statistics: mobile phones, smartphones and neither of them. Not surprisingly, the rich countries have more smartphones than the poorer countries. In some of the rich countries nearly all of the phones are smart. This raises the question whether there is a straight relationship between a country's gross domestic product (GDP) per person and the number of smartphones. This is shown in Fig. 20.5. It may be seen that there is a fairly good correlation. Wealthier countries have more smartphones.

Another question of interest is the distribution of smartphone users by age. We know from experience that older people have a rather negative approach to mobile phones. They were quite happy with those that stood quietly in the hall. Thus, it is not surprising that in the beginning of the mobile rush they were left somewhat behind. However, they could not resist the tide either. The change in smartphone penetration between 2015 and 2018 is shown in Table 20.1 for age cohorts of 18–34 and 50+. Younger people could not make much advance because they were already close to saturation. Older people, on the other hand, have made significant advances, although of course they are still lagging behind

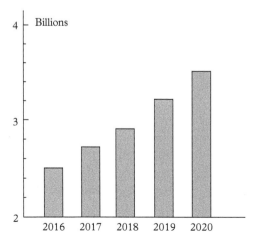

Fig. 20.3 Smartphone subscriptions from 2016 to 2020. Redrawn. Source: bankmycell.

Fig. 20.4 Percentage of adults having smartphones (yellow), mobile phones which are not smartphones (blue), and no mobile phones at all in 2018. Redrawn. Source: Pew Research Center.

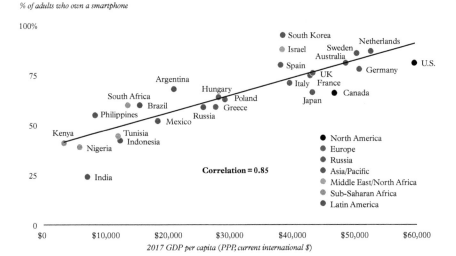

Fig. 20.5 Mobile phone penetration versus GDP. Redrawn. Source: Global Attitudes Survey, 2018, Pew Research Center.

the young ones. Canada is an exception. The penetration of both cohorts slightly declined between the years. It is difficult to guess why. Maybe the size of Canada has something to do with it. Having lots of lakes and woods, maybe some people realized that there is life beyond the smartphone. Maybe they surf the waves on the lakes instead of surfing the Internet.

Finally, there is dependence on gender. One would assume that having a mobile phone is gender neutral. Statistics show that for the first 30 years there was a slight male majority but that turned later into a slight female majority. To explain this, one would need to delve into the differences

Table 20.1 The percentage of population having mobile phones in a number of countries by the 18–24 age cohort (second column) and by the 50+ cohort (third column). The upper value is for the year 2014, the lower value for 2018.

Australia	95	58
	97	68
Canada	94	46
	90	43
France	85	22
	97	53
Germany	92	40
	98	64
Israel	87	50
	91	80
Italy	88	35
	98	48
Japan	77	18
	96	44
South Korea	100	74
	99	91
Spain	91	51
	95	60
UK	91	44
	93	60
US	92	53
	95	67

Source: Spring 2018 Global Attitudes Survey Q46. Pew Research Centre.

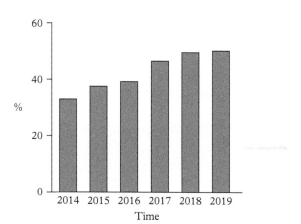

Fig. 20.6 Mobile share of the Internet, 2014–2019. Redrawn. Source: We are Social.

between the two sexes. This is highly unfashionable at the moment (2020) so I am not going to speculate.

Some time ago one needed a computer and a modem to access the Internet. Modems died a quiet death years ago, with computers now using broadband, of course. However, here again, mobile devices are on the march. In the five years between 2014 and 2019, their share as the means to access the Internet increased from 33 to 50.1 per cent, as shown in Fig. 20.6.

Problems with Statistics

Our generation not only believes in statistics but is addicted to it. This was not always the case as the quotation 'lies, damned lies, and statistics' speaks for itself. The authorship of the saying is not clear and it originates from the late nineteenth century. I like to think that it was said/written by Mark Twain first. It is the kind of thing he might have said. One might wonder which of the statistical statements made in this chapter belong to damned lies, the second category? This subject might indeed qualify. Having lots of smartphones is a status symbol; therefore, the governments of some countries might be tempted to exaggerate the number. This may happen, but the main reason is the essential ambiguity of the matter. There are several different definitions of what it means to have a mobile phone/smartphone. When we talk about mobile connections, we mean not the number of mobile devices but the number of SIM cards. An individual can have more than one SIM card; therefore, this measure very much overestimates the penetration of mobile phones. Unfortunately, when statisticians talk about mobile devices, very often they include in this category tablets, dongles, and routers. So that leads to an underestimation of the number of mobile phones.

In the developed world in a family of five, a couple and three teenage children, the likelihood is that each one has a mobile phone. So if a mobile phone per family is counted, that will underestimate the total number. In the developing world it might happen that a mobile phone is used by several people who live in the vicinity. So that underestimates the number of users. The result is that different statistics might not be directly comparable. When it is claimed that there are already more mobile phones than people on Earth, that is probably an overestimate. But does it matter if these statistics are not quite perfect? Not for the layman. It matters, however, for the companies selling mobile phones. They need accurate information for targeted marketing.

Social Impact of Mobile Phones

Let us look first at the impact of mobile phones as telephones. We know that the telephone that came in 1874 in Boston, courtesy of Alexander Graham Bell, had an impact upon the subscribers. It was a good thing for businessmen. They could discuss business transactions over the phone. That meant, on the whole, higher efficiency and higher profits. For their wives it gave an opportunity to express their opinions about third persons. That's what I wrote in Chapter 5:

[The telephone] might have made Mrs Jones an enemy of Mrs Smith on account of what she said about her to Mrs Robinson in the course of their lengthy telephone conversations but it did not in many other ways contribute to the spread of hostilities.

So both businessmen and their wives made good use of the telephone. And that good use lasted for over a century. When mobile phones came, it was an advantage for businessmen and businesswomen. It might have made them a little more harassed, to be available at any time and any place to a caller, but it had little impact on their social life. It is the emergence of smartphones that have made all the difference. It is a device that can be used for practically everything. Well, perhaps not everything. A smartphone is entirely unsuitable for boiling, for example, an egg, although even in that case if you want to consume an egg your smartphone might give you a range of recipes. The main impact for adults came with the possibility of accessing the Internet. I know people who can pull out their smartphones at the slightest provocation and look up the subject being discussed. That's good; that makes people more cultured, wiser, and perhaps even more tolerant.[1] On the whole the impact on adults have been minor and salutary. They have made good use of the facilities offered, GSM for navigation, photography, taking videos, playing games, etc., but it has not changed their lives.

It is the children born in the first decade of this century who are mostly affected; we might even say addicted. We often see bevies of girls in parks and on the streets taking selfies or just expertly clicking their smartphones. Boys are at it too but perhaps less publicly. Interestingly smartphones are loved and cherished by all children independently of age and religion (Figs. 20.7a, b, c)

Children take to smartphones as if they were born with them. Some of them even take their smartphones to bed, checking them for the last time before falling asleep. Why such addiction? Had the practitioners of psychotherapy known about it? Obviously not. Had they been able to predict the mass psychosis the companies making the smartphones would have brought out their models earlier. In fact, there is no plausible explanation even after the event. The emphasis is on the word plausible because there is no shortage of explanations. Explanations, in fact, abound but there is just no consensus. I could offer one explanation myself. I think it is the same kind of relationship that keeps cults together. The relationship between the leaders and the led. For the majority of children it is a relationship between the bullies and the bullied. There have, of course, always been bullying in school. The bullies have been

[1] Seeing more of the world, understanding more of the world, would probably mean a better understanding of people as well.

Fig. 20.7 (a) Pre-teen, (b) teenage, and (c) Muslim girls using their mobile phones. Source: Shutterstock.

few, the bullied many. The bullies have had strong personalities; the bullied have been meek. No doubt, this is still the case but there are two differences. In the old times bullying was restricted to school time and it was verbal. As the Romans said, 'verba volant scripta manent'.[2] In the old times after school a child was free and could shake off any unpleasantness. Now, messages appear on his/her smartphone at any time of the day that require some kind of response. The stronger ones among the bullied will erase an offending message. The weaker ones will brood over the words received. They might even look at them again and again in the safety of their own rooms.

[2] 'The spoken word flies away, the written word remains.'

Cartoons

After the seriousness of bullying let us employ a lighter touch and look at some cartoons. First about adults. Everybody nowadays is so much used to holding and looking intensely at their smartphones that even when some major event occurs (e.g. they die and go to heaven) that deprives them of their phone, they keep on staring at the spot where the phone used to be (Fig 20.8a). Young boys are also affected. They used to go to the park to play, football being the most likely game. Nowadays the young boy wanting to play football is an oddity. The great majority are happy to click buttons on a smartphone (Fig. 20.8b).

Fig. 20.8 (a) Mobile phones in heaven; (b) the decline and fall of football. Source: Digital Synopsis.

Fig. 20.8 (b)

(b)

CHAPTER TWENTY-ONE

Artificial Intelligence

Introduction

Why write about artificial intelligence in a book devoted to communications? I feel it is an example of collaboration between two disciplines. We might call it, as it sounds much better, the intelligence–computer symbiosis. It is not in a much advanced stage at the moment. Some people claim that the present state of artificial intelligence is just a little short of that of a lowly worm. Nevertheless, it is worth a short chapter and it is of further interest in Chapter 22 concerned with the future.

Artificial intelligence originates from the combination of two different subjects: the art of creating intelligence and installing automation. The hope is that as time goes on machines will have more intelligence. In general, the best way to create intelligence is to create humans. In ancient Greek mythology this was indeed done. The creator of man was Prometheus who belonged to the Titans. His material was clay. The first woman, Pandora, was created independently by Zeus from unknown material. Somewhat later and possibly before, the Judeo-Christian god created first a man called Adam from dust and then a woman, called Eve, using some spare part from the already existing Adam. In modern times the creation of humans was deemed possible, at least in fiction. According to Mary Shelley, the capable Dr Frankenstein, a scientist very much on the mad side, created a man using body parts from various sources.[1] It was not a happy creation. The creature turned out to be ugly as a monster and was regarded as such by all and sundry. The monster was not happy either. He felt lonely and wanted a wife, which Frankenstein failed to provide. In revenge, the monster committed a few murders that included Frankenstein's wife. After that one could not expect a happy ending. And there wasn't one either. As the novel ends both the creator and the creature die.

Leaving mythology and fiction behind we can look at attempts to realize artificial intelligence in our time. First, how can we define it? Let us see some definitions:

A branch of computer science dealing with the simulation of intelligent behaviour in computers.

Or one that is practically equivalent:

The capability of a machine to imitate intelligent human behaviour.

[1] Her novel was entitled *Frankenstein or the modern Prometheus.*

Getting the Message: A History of Communications. Second Edition. Laszlo Solymar, Oxford University Press (2021).
© Laszlo Solymar. DOI: 10.1093/oso/9780198863007.003.0021

Or another similar one:

The capability to build systems that think exactly like humans do.

Or a more elaborate definition:

Artificial intelligence is the theory and development of computer systems able to perform tasks normally requiring human intelligence, such as visual perception, speech recognition, decision-making, and translation between languages.

Assume we have a machine properly endowed with artificial intelligence. How can we ascertain that the machine created to be intelligent can really match human intelligence? There is a test devised by Alan Turing in 1950 known as the Turing test.[2] The relevant experiment is done by having a human and the machine, both out of sight, waiting to be interrogated. The interrogator asks a number of questions from both. Can he/she tell from the answers received which of the two is the human and which is the machine? In my view, if the test is done casually, if the range of the subjects covered is narrow, if the duration of the test is short, then it is quite likely that the two cannot be distinguished. However, for a sufficiently broad line of interrogation, conducted for a sufficiently long time I feel it is possible to tell them apart—perhaps not for ever. It may not be the case three decades hence.

The other component is automation. Some examples of that can also be found in the works of early Greek authors. In Homer's *Iliad* the gates of heaven are opened automatically to allow the delivery of some highly needed nectar for the gods' banquet. Delivery is by driverless chariots. These are good examples showing what automation can ideally provide but not how it is done. A simple example of automation will help in understanding the basics of what makes a process automatic.

First, a familiar household example. An apartment is heated by a central heating system. The aim is to keep the apartment at a constant temperature, T. We must then have a sensor, a thermometer, that measures the actual temperature, t. If t is less than T, then more heating is required. A feedback process then switches on the boiler of the central heating system. When the required temperature is reached, the feedback system acts to turn off the central heating.

A second example is taken from the beginning of the Industrial Revolution, the control of a steam engine. It needed some control because if the steam engine run too fast, there was a danger that the machine will be ruined. So the speed had to be limited. It could be done by a human watching the machine. If he saw the engine running too fast, he restricted the flow of steam by moving a valve. James Watt replaced the human with an ingenious machine known as the fly-ball governor. If the speed was too high, the governor moved upwards,

[2] Turing was a brilliant computer scientist, the head of a team that broke the German Enigma code during the Second World War.

causing a movement that restricted the steam flow and slowed down the machine. It was one of the first automatic devices of the modern age. It had something that all automatic devices have nowadays, an error-detecting mechanism and feedback. Is that artificial intelligence? No, because it is a deterministic device. Everything is decided in advance. It is known how the position of the governor is related to the amount of steam entering the engine.

In order to make clearer the difference between automation and artificial intelligence we shall discuss the computer programming of two games, Hex and draughts, both played by two persons.[3] It has been shown by mathematicians, by looking at all the possible moves of the two players on a 7×7 Hex board, that the first player can always win. This I would call automation without any hint of intelligence because the computer is programmed so that it is able to specify in advance all the required moves of the first player in response to any possible move of the second player,

Now let us look at draughts. It is a game much better known than Hex because it is played on a chessboard. A computer engineer Arthur Samuel rose to fame as the man who wrote a computer program for playing draughts. The solution offered for playing Hex on a 7×7 hex board was not available to him because the computer he used did not have sufficient memory to store all the possible moves. Instead he developed an alternative approach. At any stage of playing the game there are N possible moves. Samuel developed a scoring function that attached a score to every one of these N moves. His function took into account the number of pieces each player had, the number of kings, and the number of pieces with a chance to become king. In addition, he improved his scoring function by making the machine learn from the large number of games it played. Is this an example of artificial intelligence? To choose the next move the computer had to find which move gave the highest score. It was all pre-programmed; the computer did not rely on its own intelligence but on the intelligence of the programmer. Nonetheless, I would regard this as an early example of artificial intelligence because the machine could learn from the games it played earlier.

When Samuel's computer program played successfully against human draughts players, it gave satisfaction not only to him but also to the company upon whose computer the program worked. The president of IBM, Thomas J. Watson Sr, predicted that the demonstration of playing draughts will raise the price of IBM stock because there is now a new set of people who will buy IBM computers not for business, not for scientific calculations, but for entertainment, for playing games. How right he was!

By 1997 IBM's supercomputer was ready to challenge Gary Kasparov, the world chess champion, to a match of six games. The idea was to do the same kind of programming that was used by Samuel: to devise a

[3] Hex is a game played by two players on a rhombus-shaped board containing $n \times n$ holes. The aim of player A is to establish a contiguous line in the direction of one of the axes of the rhombus, whereas player B wants a contiguous line along the other axis. This is to be achieved by inserting pegs into the holes. One of the players is bound to win. A draw is not a possibility.

score function and to play always the move that gives the highest score, but the difference was the number-crunching ability of the supercomputer. It could analyse 200 million moves per second. At the end Kasparov narrowly lost. He claimed that the IBM team cheated. Some of the moves were played by humans and not by the computer and that is cheating because he would play differently against humans than against computers. The controversy was never resolved. Anyway, whether humans provided a helping hand or not, this was an amazing feat. And, of course, it had the advantage that the layman could understand what was achieved.

Robots

Again, antiquity has the priority. Hesiod in sixth century BC wrote a poem about a robot known as Talos, programmed by the god Hephaestus to defend Crete against pirates and all other enemies.[4] It was very tall with humanlike features, with a presumably grim expression, and made of bronze. Unfortunately, it was destroyed by the Argonauts on their way to Colchis to get the Golden Fleece.

The modern version saw the light of day in 1921 in a play by the Czech playwright Karel Capek. The title is *R.U.R.*, an acronym for Rossum's Universal Robots.[5] This was the first time that the term robot was introduced in the context we still use today, an artificially created human bearing close resemblance to genuine humans produced in the traditional way. It is a dystopian story. It starts well, the robots serve the human race faithfully, but then, realizing their power, they revolt and kill all the humans but one.

An even more modern version is offered by the 1982 film *Blade Runner*. The robots there are called replicants. The story is essentially the same but more sophisticated. The replicants revolt. Some of them kidnap a space ship and return to Earth with no good intentions in mind. The scene is a futuristic Los Angeles where a retired policeman has the job to hunt them down.

Genuine robots, I mean in real life, not in fiction, exist also. They do all kind of jobs for humans. The country which uses them most is Japan. There might be two reasons for it. Firstly, the Japanese believe that all objects have a soul, so robots are more accepted in their society, and secondly, Japan's population is shrinking and there is need for additional work force. A rough estimate of the number of robots in Japan is a quarter of a million. Some of them are quite sophisticated. Mindar (Fig. 21.1) teaches the tenets of Buddhism in the Kodai-ji Temple in Kyoto. There are also some hotels in Japan staffed by robots.

In the West we live in a secular age. Despite that, robots that look exactly like humans, e.g. Mindar, make us uncomfortable. They are graven images which according to the Ten Commandments we should not make for ourselves.

Fig. 21.1 Mindar is very human-like; it is worth looking at the skin and nails. Source: Getty Images.

Some Potential Applications of Artificial Intelligence

We shall list here, in no particular order, a number of applications of artificial intelligence which already exist or are just around the corner.

Process automation

This is an extension of automation to administrative processes. Whatever automation did for manufacturing, it could do the same for administrative processes, e.g. writing a report when all the necessary data are available in the computer memory or choosing a shortlist from a large number of applicants for a particular job. I would regard the former application as belonging to genuine artificial intelligence, because apart from the need to understand the significance of various factors it must also be able to express the conclusions in natural language and, even more importantly, learn that all the negative results are due to external factors beyond the reach of the company. I would doubt though that any CEO would be willing to release a report written in such a way. In the latter case of the shortlist, the program would look at keywords, use a weighting function for all of them, and recommend those with high scores to be interviewed. I would be more willing to rely on this because humans are erratic when making quick decisions and machines might indeed do better.

Personalization

Let us first see some definitions.

1.   Personalization is about the dignity and well-being of the individual.

2.   Personalization restores the individual to prominence.

3.   Personalization is an approach to social care that focuses on putting individuals at the very centre of the support and services they receive.

4.   Personalization is the art of offering tailored marketing.

Numbers 1 and 2 sound like fashionable overstatements meant for those who are likely to believe it. Number 3 sounds a little more realistic. There is no doubt that in health care personalization is necessary but it is not a new idea. It had been probably practised since the time of Hippocrates. A new trend is epitomized by number 4. All companies in the retail business make desperate efforts to learn more about the likes and dislikes of customers, and then send them targeted advertisements.

Those who produce the programs regard this as an application of artificial intelligence, but I don't see that much intelligence is needed to choose items for recommendations once the shopping habits of people are available. I would regard this as a good example of automation. Unfortunately, this is not an innocent business. Once information about an individual is available, it can be easily misused by criminals and dictatorial governments.

Creating music, novels, plays, and poems

It must be pretty difficult to write the programs but only experts can comment on their originality or otherwise. Listening to or looking at any particular composition, I am sure there is never a consensus. What one regards as good or bad is entirely subjective. But of course this also applies to all the various activities in the business world.

Virtual assistants

There are several available: Apple's Siri, Google's Assistant, Microsoft's Cortana, Amazon's Alexa. The feat they can perform with various success is recognition of natural language. The point is not whether they can give the correct answer to a question (that depends on what information they can access), but whether they understand the question. I regard these as excellent examples of artificial intelligence, but they need a lot of improvement before they can become really useful.

Cybersecurity

Artificial intelligence might be useful in defending against the cyberattacks created by the existence of cybernetics. One direction in which some progress has been made is fraud detection. From many, many examples the algorithm learns what is normal activity and detects deviations from there.

Driverless cars

The applications described so far have not excited much interest, apart from the companies which sell the software. On the other hand, driverless cars (also known as self-driving cars and autonomous cars) are not only talked about but can be actually bought if you can afford them. The reason is obvious. We would all love to have cars which can provide the same service as taxis, taking you from door to door. We would all love to enter an address into some device on the instrument panel, and then settle with our favourite newspaper on a comfortable back seat and not to budge until the destination is reached. I would regard this as the best test of artificial intelligence. There are so many possibilities to look at, and there is so much at stake. At present none of the various makes are ready to offer that driverless experience. They can do much less. The person responsible for the car is advised to sit in the traditional driver's seat and be ready to take over the wheel if needed. When will they be ready? In ten years? Thirty years? Fifty years? Nobody knows.

I think a serious obstacle for reaching that stage is the fear of accident. We can accept that car accidents occur. Those of us interested in statistics know also that there are about 1700 fatal accidents in the UK each year, a very large number. It is quite conceivable that driverless cars will make fewer accidents but psychologically we are less ready to accept them. We accept human error because we often do that ourselves. But if there was a fatal accident caused by a driverless car we would blame the manufacturers and demand that their product be withdrawn from the roads. So my guess would be 50 rather than 30 years.

Writing private letters

This would be a good area for artificial intelligence to proceed. The demand is there. Children of the Second Industrial Revolution find that writing a witty letter demands too much brain power. So they would be quite happy to buy a software package if available. Some rudimentary programs already exist. If one wishes to write a letter to a friend who is going to retire, such help is already to hand. Having many friends reaching the retiring age I did some research and found the following letter offered by a word processing software:

Dear John,

Face it. The time has come when you're going to have to confront some really difficult decisions. Whether to wake up early or sleep in. Whether to go sailing or hiking. Whether to paint a picture or a room. Whether to watch a movie or a sunset. You've got some tough choices ahead. I hope you thoroughly enjoy every one of them. Congratulations on your retirement.

In case I do not want to send the letter exactly in the same form, my program also offers some useful tips: 'Convey a sense of unlimited options to the reader', and 'a little humour gets the message across'.

Given another 15 years, programs will be able to do a lot more. I suppose by then I shall be able to type in keywords like friend, pompous idiot, no sense of humour, proud of his prowess in golf, just married for the third time, retires soon. And then I would jerk the mouse, bash a couple of keys, and a letter of sufficient civility and solemnity will emerge in a couple of minutes.

Medical diagnoses

This is an ideal subject for artificial intelligence. A properly programmed machine is better at it and arrives at conclusion in a fraction of a second. Proper programming means that a very large number of cases are entered into the machine's memory. They are particularly good at

Fig. 21.2 "You are probably wondering how I got the nickname kill-bot. Well, it's a funny story." Source: Cartoon Collections.

"You're probably wondering how I got the nick name, kill-bot. Well, it's a funny story..."

analysing MRI and X-ray images, including CT scans, but they can offer significant help to doctors in most branches of diagnostics as well.

Medical operations

Too early for robots to conduct medical operations, the surgeon remains in control but relevant additional information could be offered in real time during the actual operation by machines.

So is artificial intelligence used in medicine? Not much at the moment but it is probably the field with the highest potential. Even small steps may offer significant benefits.

Let's finish this chapter on a lighter tone. According to Figure 21.2 there are dangers in trying to let robots into cocktail parties. Figure 21.3 shows that the implications of the Turing test have reached wider sections of society.

Fig. 21.3 "I won't date another AI engineer unless he can pass the Turing test." Source: Cartoon Collections.

"I won't date another AI engineer unless he can pass the Turing test."

CHAPTER TWENTY-TWO

The Future

Technical Predictions

The easiest prediction is a simple extrapolation from existing trends. If the number of smartphone subscribers has been increasing over the past 10 years, then it is fairly safe to predict that it will keep on increasing for a while. At what rate? At the rate it increased in the past year.

Can one make even better predictions? Mathematicians will say that it is necessary to look at the higher derivatives; i.e. one needs to know not only the rate of change but also how over the course of the past year the rate of change has changed, etc. This is the way professional forecasters, say those employed by telephone companies, are bound to proceed. At the same time their marketing people investigate the likely demand for some new services they intend to introduce, and then they extrapolate from the various surveys conducted. This kind of prediction envisages relatively slow change.

Once in a while, some people might make quite imaginative leaps into the future. For example, a *Punch* cartoonist expressed the view in 1879 (Fig. 22.1) that it would be possible in the future not only to talk to people in faraway countries, but to see them as well. A French prediction also dating from the nineteenth century is shown in Fig. 22.2. Interestingly, neither cartoonist could make the further mental jump of dispensing with the wire. The possibility of picking up signals from thin air was too radical even for the most imaginative. But once radio was accepted as a reliable means of communications it was not particularly difficult to prophesy that more and more ships would be supplied with radios, that radio messages would be sent farther and farther until one day it would be practicable to establish radio contacts between any two points on Earth.

What would be my prediction for the next ten years? More of the same thing. More satellites, more optical fibres installed, higher capacity for all optical cables, the demise of simple mobile phones, smartphones for everybody, more data communications, fantastic bandwidths making possible instant downloading of movies, wider access to the Internet, more services on the Internet, bigger computers for scientists, smaller computers for businessmen, and the price of all these will keep on falling and falling.

Am I suggesting that there are no longer any radically new devices to be engineered and introduced on the market? Not at all. But if a new device does not fit well in the existing pattern of development, its

Getting the Message: A History of Communications. Second Edition. Laszlo Solymar, Oxford University Press (2021).
© Laszlo Solymar. DOI: 10.1093/oso/9780198863007.003.0022

Fig. 22.1 Mr Punch predicted conference television a century and a half ago. It is remarkable that this cartoon in *Punch* magazine for December 1879 foresaw not only UK–Australia conference television, but also large-screen projection in a format not dissimilar to today's high-definition TV. Source: *Punch*, 1879.

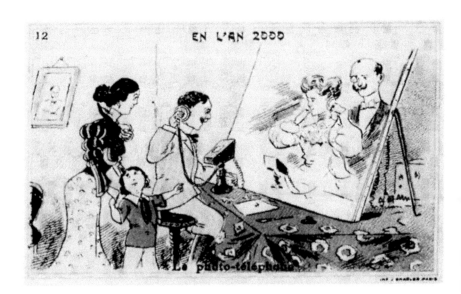

Fig. 22.2 A French prediction for the year 2000. Source: Archives, French Telecom.

chances are more limited. Take three-dimensional imaging, for example. The principles were established by Dennis Gabor's invention of holography more than half a century ago. The ultimate videophone would be a device in which the full three-dimensional image of the person at the other end of the line would appear. I mean a genuine three-dimensional image so that one could admire a new dress or a new hairstyle by going round the image suspended in air. I believe such a system could be developed in 15 to 20 years but I think it would be too expensive. I think some cheap imitations of three-dimensional effects will be available but they would not be good enough for commercial success. The added pleasure of seeing things in quasi-three-dimension would not be adequate compensation for the price that would still be too high.

What are the kinds of technological changes of which one can be reasonably certain? Those where the principles of operation are all known and where the engineering problems may be expected to be ironed out without too many difficulties. Let me describe a few.

Semiconductor devices

Will Moore's Law continue to be obeyed? Well, perhaps the rate of increase in the number of semiconductor switches on a single chip will slow down, possibly doubling every three or four years, but I have no doubts that there is still a lot of potential for further expansion of the known technology. Improvements can be expected in four directions: (i) growing integrated circuits in the third dimension (instead of using merely a single surface); (ii) using other materials than silicon (in which electrons can move faster); (iii) moving towards shorter electromagnetic waves for producing the circuits (thereby reducing the dimension of a single device); and (iv) changing to new types of devices (e.g. ones based on resonant quantum mechanical tunnelling).

Superconductivity

It is an extraordinary phenomenon: many substances are capable of losing all electric resistance below a certain critical temperature. It was discovered a good hundred years ago. For many decades progress was very slow, then in the 1980s due to new discoveries there was a sudden advance: the critical temperature increased by a factor of 4 but it is still very low. It is around −70°C. The physical mechanism is still unknown. I somehow feel that it would be very unfair of physics to put such an important phenomenon into a dark corner. I feel, a mere guess, that within the next two decades room temperature superconductivity will be discovered and with that a drastic reduction in the price of electricity, a drastic reduction because losses, inevitable now, will disappear.

Display devices

The workhorse of all display devices has been the cathode ray tube ever since its invention by Karl Ferdinand Braun over a century ago. These have disappeared completely and young people today would not know how bulky they were. Display devices based on plasma effects have been with us for longer than a decade, but their manufacture has recently stopped. Light emitting diodes made of silicon have also been widely used: they are also on their way out. The new champions are light emitting diodes made of organic material, abbreviated as OLEDs, which provide wider viewing angles and better image quality. They are more expensive too. In a decade they will dominate the market. At the end of my 10-year period, I predict the standard display device will hang on a wall and have dimensions of 1×2 metres.

So far I have assumed that our present habits of looking at displays will remain unchanged. In the future will we want to watch a screen far away? After all, if we wish to listen to music we may elect to use a headset which shields us from external noises. Why not do the same for images? Could we not have a headset which projects the required information right onto the retina? That would solve the problem of strong ambient light falling on the screen and, besides, it would be more private. Such systems are already in existence in the laboratory and, I expect, will be widely available within 10 years.

Satellites

Will satellites have increased significance in the future? There is no doubt that they are needed for global coverage, so there will be more and more of them with signals travelling between them, or they might be used only as transponders, connecting one continent to another one.

Balloons

The idea has been there for quite some time, I mean the idea of lighter-than-air vehicles. The problem of how to keep them up in the sky was solved more than two centuries ago by the Montgolfier brothers. For more serious technical applications we tend to call them airships nowadays. They stay in place as the Earth rotates because the atmosphere drags them along. In practice of course they would not stay exactly overhead because the wind would blow them away. The effect of wind can be minimized by choosing a height at which winds are relatively gentle and, besides, the position can always be corrected by a small engine at a low consumption of power. The advantage of airships is that they are not too expensive to build and can be fairly easily put in position. They can also be very big, capable of accommodating lots of equipment on board. They could easily provide mobile two-way communications within an area of, say, 500 kilometres across. Their main disadvantage, at least for the moment, is that they require a big ground-support team, but with improved control systems that could probably be reduced in the future. Being big in size has the further disadvantage of offering an easy target even for not-too-well-equipped terrorists. So, considering all the cons and pros, will airships come? They might.

Optical fibres

Will optical fibre come to every home within the next 10 years? Well, there would not be much point to take a fibre to a single subscriber in a hut in the middle of the Sahara or to an igloo close to the North Pole. But for densely populated areas, yes, my guess is that optical fibres will come to every home.

Speech recognition

The idea that inanimate objects can recognize the human voice was established in most of us at a tender age. Let us remember the Wicked Queen's question:

Mirror, mirror on the wall,
Who is the fairest of them all?

The mirror if you remember, fully comprehends the question, scans the beauties of the land, and provides an immediate answer as to who is the fairest of them all. Well, if a mirror can do it, so can we. After all the principles are really simple. They are closely related to those of the telephone. Voice carried by an electric current has a certain pattern which depends on the sounds uttered. Hence a proper analysis of the pattern can lead to recognition of speech. This is quite an old science. By now speech recognition systems are commercially available and used in every walk of life. We should not forget that in the Snow White story it is not a simple speech recognition that is performed. The mirror also understands the command it receives. It has a job of scanning all the beauties of the world. Today, this is done by virtual assistants as mentioned in Chapter 21.

Recognition of handwriting

The technique has been around for some time as part of the problem of pattern recognition. It is quite easy to recognize letters when they stand apart, but to analyse written text is much more difficult. The best technique is to put software into a pen which recognizes the letters immediately as they are written. There is no doubt that future software will be much improved and will diligently learn to read many different styles but interest in this field of study is bound to decline as fewer and fewer people practice the art of handwriting.

Other possibilities

What other possibilities are there for the transmission of information? What we call multimedia is really restricted to sound, text, image and video. There have been no attempts so far to transmit smell, taste, or touch. Could it be done? I think, yes, provided sufficient effort is invested. I see a possibility that the problem of smell transmission could be solved within 10 years or so. It is after all possible nowadays both to analyse and to synthesize smell. This is what all the major perfumeries do. The question is again whether it would be worth the effort. I think taste and touch are in a comparable situation. Society might be willing to pay for the blind to see and for the deaf to hear but not for transmitting smell, taste, and touch. And besides, would not a medical engineer be much prouder if he/she worked on giving sight to the blind rather than on transmitting the taste of chocolate?

What about entirely new means for long distance communications? Could matter waves be used for sending messages? There has been definite progress in this field ever since de Broglie postulated wave–matter duality in the middle of the 1920s. There are now so-called atomic lasers available which can produce pure matter waves. I do not think though that they could be harnessed for communications purposes. Or let us make an even wilder assumption. Will anything be discovered that would show that our Einsteinian view of the Universe is not entirely correct and that communications at speeds faster than that of light is possible? It would certainly be a good thing to avoid the irritating delays in telephone conversations that occur when geostationary satellites are used but, again, I would not be optimistic.

Let us finish this section with one of the techniques that already exists but feels like science fiction: interaction between the brain and soulless devices like computers or cars, known as neural interfaces or brain–computer interfaces. These work by using electric brain signals as transmitters while the receivers rely on traditional electronics. It has been proven experimentally that thought can indeed make light model vehicles move, but it is far from clear what exactly are the thoughts responsible for the driving. Maybe the only connection that exists is that once the brain produces electric signals those signals can be used for any purpose including moving light vehicles.

A phenomenon even closer to science fiction is brain-to-brain communications. Experiments have been done on both animals and humans which claim successful transfer of thought. So telepathy exists. Luckily, all the efforts so far have been very rudimentary. Beware, this is an extremely dangerous field, the playing field of dictators. Efficient brain washing comes after that.

The Social and Economic Dimension

How will life in the West change as we become more and more immersed in the communications revolution? How will our businesses be run? What are our economic prospects? These questions and many others have been analysed in detail by Frances Cairncross in her book *The Death of Distance*. Anyone wishing to delve into these problems is well advised to read it. To my mind it is a little on the optimistic side, but I hope she is right. In what follows I shall be unable to match her analysis in any way. I shall restrict myself to a few specific topics.

Economics is not a field into which I would dare venture. I would just like to mention one single aspect, the speed with which information can now reach all relevant parties. Might it cause havoc? What would happen now if some scenarios from the past were replayed? Let me take as an example the devaluation of sterling in 1967. The British currency had been under pressure on and off during that year. On the 16th November James Callaghan, the Chancellor of the Exchequer, was asked in Parliament whether reports in the press and television on negotiations for a 1 billion dollar support loan were correct. His reply gave the impression that the stories were untrue. The markets immediately assumed that the other alternative, namely devaluation, had been put on the agenda. There was a run on sterling on 17 November. Devaluation was announced at 9.30 p.m. on Saturday, 18 November.

With today's communication facilities and considering the ease with which currency dealers can now buy and sell their ware, the run on sterling would have taken on tragic proportions within the hour. Engineers are well aware of these problems. There is always some chance that in some plant, some variable goes out of control, and the task is to bring it under control. The time available to make amends may not be more than a fraction of a second. The discipline concerned with these problems is called control engineering. It is a sophisticated science. One hopes that the economists at the Treasury (and not only at the UK's Treasury) have been willing to learn from the engineering profession and are well prepared to tackle such challenges.

I also believe in the benefits of horizontal integration. It means that people with the same interests can contact each other and can develop and promote further friendships, irrespective of physical distance. If somebody's interest is in, say, a rare tropical plant which he tries to grow in a moderate climate, then it would be a considerable advantage to the person to talk to like-minded people all over the world. I do realize, however, that these kinds of activities would be minor concerns. A much likelier scenario is a deep interest in drinking beer until one becomes senseless. There is then no need to search for soul mates far away. The chances are quite good that a suitable companion may be found in the same locality.

The point I wish to emphasize is that people will need to think less and less. The younger generations will take information technology for granted and will get used to it as instinctively as we drive a car today, or switch to different channels on a television set. The young will grow up with all the new gadgets and will be able to use them with skill but the question 'Why?' will occur to them less and less. This tendency will be further aggravated by the intellectual decline in our education system. We are already 'customer oriented'. Even in higher education the tendency is to attract more and more students (after all revenue depends on them) and present them with fewer and fewer intellectual difficulties. Even in Oxford, which is deemed to be one of the better universities in the world, we tend to teach less and less of the fundamentals.

If I want to be optimistic I can envisage, by the middle of the twenty-first century, a society which consists of a narrow elite and a grey mass of people who do not suffer any wants. Since work will be optional the great majority will not have any gainful (or wasteful) occupation. Their interests will be focused on playing games. They will derive their pleasures from virtual reality.[1] That may very well be a society in stable equilibrium.

Political Sphere

The advent of instant communications in the form of the electric telegraph made a tremendous difference to the world of diplomacy and politics. Oddly enough, the telephone made relatively little difference when it arrived. In modern times it is, of course, not unusual for heads of governments to call each other, and they don't even need a crisis for doing so. The main reason for more contacts at the top is not that the new means of communications have made the technical realization of such contacts easier. Had there been the will, the technical difficulties could have been overcome. I think it is rather the shrinking size of the world, the interdependence of the various states upon each other, and the new tendency of obligatory chumminess with everybody (including your fellow head of government) that is responsible for the wish to have a chat on the phone. Video conferencing by leaders at the European level will come in due course, but not, I think, for a decade or two.

Will the democratic structure further develop as a result of new facilities? The time will soon come when all adults could be provided with a voting box of their own so what's the point of having elected representatives? Why do we need a House of Commons or Representatives at all?

[1] Not unlike in Huxley's *Brave New World* in which everyone can count on *soma*, a harmless drug that will induce pleasurable dreams.

Could we not proceed to direct democracy as in Pericles' Athens? Whenever a decision is needed (e.g. should we send troops to such-and-such a part of the world, or should we reduce income tax to 10 per cent), it would be possible to ask the people. The question would appear on the voting box screen and the voters would simply press either the YES button or the NO button. I feel though that such a way of conducting the affairs of state would not be entirely free of contradictions. The electorate might endorse controversial proposals, e.g. that everyone should be able to command an above-average pay packet.

Or take another possibility that would lead in the opposite direction. The voting boxes could introduce some weighting. I remember this was proposed by a friend of mine a good 40 years ago. Everyone's vote would somehow be weighted by his or her IQ. The votes of those with higher intelligence would weigh more than those cast by people at the bottom of the IQ scale.[2] Will that come? I doubt it. Any suspicion that intelligence might play a role in the conduct of affairs will make people mount the barricades. So will there be any radical reforms? I doubt it.

Now let me look at a wider problem. Is there a relationship between the state of communications and the political structure? One possible view is that science and engineering are value-free. Their achievements can be used for good or evil. In a political structure, like that of Orwell's Big Brother, information technology is the instrument by which all human activity is controlled. Conversely, in a liberal democracy, the same means may reinforce the individual's freedom of choice. So it all depends how information technology is used—or does it?

I strongly believe that increased communication facilities were responsible for the political changes in Eastern Europe, as discussed in Chapter 16. Will improved communications lead to the collapse of communism in Asia too? There are arguments in favour of that view. Susan Lawrence, for example, reported how successfully Chinese dissidents have kept in touch with each other with the aid of pagers and how they switch to another pager before the authorities have track them down. So with the communications system steadily growing it would be difficult to silence he opposition. However, I am not so sure that what happened in Europe is a good guide for China. European communism was built on the ideas of universal brotherhood. The Communist Manifesto of 1848 tried to do for the Fourth Estate what various pamphlets preceding the French Revolution did for the Third. Of course, communism developed into something not foreseen by the founding fathers, mainly as a result of the activities of an obscure seminarist from Georgia, Joseph Stalin. He managed to incorporate communist ideas into a framework which would have been the envy of any despot. Nonetheless, the European communist movement (whether in Russia or elsewhere) never abandoned the pretence that it was the champion of working-class interests. With improved communications it soon became obvious that

[2] The basic idea came from John Stuart Mills.

the working class in the West not only enjoyed more political freedom but had a much higher living standard as well. Communism lost its *raison d'être*. It had to disappear.

It was different in Asia where the working class hardly existed at the time of the communist takeover. Their ideology, if anything, was anti-colonialist. They were, of course, quite willing to mix in some socialist ideas but those ideas were half-baked, and despite frequent revisions, remained half-baked. Once in power, all they wanted was to stay in power. Eventually though they realized that economic liberalization can lead to wealth production on a much larger scale, and they were willing to experiment with it. After a while they reached a stage when they could tell their citizens that they knew best the path to prosperity. The irony is that they might even be right, that the large majority of the citizens are happy with their economic lot. In countries in which the taming of power has no tradition, political freedom may not be high up on the agenda of the citizens.

Would improved communications change the situation? There is a small chance that it would if it was freely available. But, of course, communications are far from being free. Partly there is censorship; the Internet is heavily censored. Secondly, China has such clout in the commercial world that private owners of satellite broadcasting chains would think twice before they did anything against the wishes of the Chinese authorities.

Now let me come back to the plight of the West and let me try to be a little pessimistic. Will we ever reach that admirable equilibrium in which everyone, both the haves (who have the secrets of technology at their fingertips) and the have-nots (who have nothing at their fingertips), will be happy? The road to salvation might turn out to be impassable because of giant potholes suddenly appearing. I can think of a few.

Political instability in the West. As a consequence of polarization there will be some small, inspired, closely knit, and determined organizations preaching the liberation of the have-nots against the haves. The have-nots will be reluctant to rise because they have all the material comforts. But let us assume that a major disaster occurs due to some software failure. The have-nots will feel betrayed. The small organizations will say, 'I told you so, go and destroy your gadgets! Have-nots of the world unite!' And the call for destruction might succeed when the next software failure causes some major disruption.

Takeover by organized crime. We might look at the present and future states of communication networks as a battleground between good hackers and bad hackers. What if one day the good hackers lose motivation and the bad hackers triumph? What if one day all the banks in the United States find their accounts cleanly erased, and the bad hackers claim responsibility? The perpetrators could then claim that they can cause damage to any part of the world by activating their programs hidden in the software jungle of various military and commercial organizations. They could hold the world to ransom. They could give a warning that any attempt at countermeasure by the good hackers will lead to immediate retaliation. And to back up their threat they could, when attempts are discovered, disable, say, half a dozen major airports for the greater part of a day. And from then on, they could rule the world for ever and ever or at least until they got tired of it.

The triumph of religion. Some authorities in some countries may decide to save the world from the satanic practices of the godless information technologists. In order to destroy that technology they would have to learn it first. So they send their young, talented, and properly inoculated men and women to study computer science at the better universities in the Western world. When those young people get their PhDs and return to their native land, they practise further the art of hacking, with all the resources of the state behind them. When they are ready, they are given the job of penetrating the major military establishments of the Western countries. And when they succeed in doing so, they give instructions for American planes to drop nuclear bombs upon their own cities. And the godless will cease to exist for the greater glory of God.

Are these scenarios likely? They are a little crude but not impossible. The future, even disregarding the biological time bomb, does not seem that certain. The past is an indication that humankind has always misused its powers. However, there are some hopeful counterindications. In the seven decades since the end of the Second World War, most of the world has been at peace. Maybe there is a fair chance that the marvels of modern technology will be used for good and not for evil.

The Mechanical Telegraph in Other Countries

Sweden. The Swedes were probably the first Europeans to follow Claude Chappe's lead. They started their experiments as early as October 1794. Several lines were set up, and the one around Stockholm played a significant role in the war against Russia in 1808–9. The Swedish mechanical telegraph was the longest to survive in Europe. It still operated in 1881. The reason for the delay was the difficulty in laying underground cables between the myriads of islands in the Stockholm archipelago.

Denmark. The Danes were only a couple of months behind the Swedes in starting experiments but they were rather slow in actually building stations. At the time of the Battle of Copenhagen (1801, when Nelson turned a blind eye to the telescope) only two stations were operational. Perhaps the most important feature of the Danish system was the link between Jutland and Funen and between Funen and Sealand. Those straits were often controlled by the British Navy, so the only reliable means of communicating between them was by telegraph.

Norway. Norway belonged to the Danish crown at the time. Their long coastline was vulnerable to attacks by the British Navy. They set up an early warning system consisting, at its peak, of 175 stations stretched over 1300 km.

Germany. The lack of a unified German state made the setting up of telegraph lines a low priority. Construction of a longer line was authorized only as late as 1832. When it opened in 1834 it stretched from Berlin to Koblenz (covering about 600 km with 60 stations) with an extension to Aachen.

Spain. Although telegraphs are known to have existed around Cadiz during the Napoleonic wars their later development was rather slow due to the numerous political upheavals in the country. A Madrid–Irun line to the French border was inaugurated in 1846. It then took no more than 6 hours for a message from Paris to reach Madrid.

Portugal. A line of 340 km with 25 stations between Lisbon and Oporto was established in the 1830s. Apparently, there was also a line from Lisbon to Badajoz which joined the Spanish system around 1850.

Russia. Lines of strictly military significance were built from St Petersburg to Petrokrepost in 1824 and to Kronstadt (an important naval base) in 1834. A much longer line (830 km comprising 148 stations) between St Petersburg and Warsaw was completed in 1839, presumably with the aim of governing Russian-occupied Poland a bit more efficiently. In Finland, which also belonged to Russia at the time, a line was built during the Crimean War (1854–6) from St Petersburg to Hango (the westernmost tip of Finland in the Baltic Sea, which came into the news again in 1939) to report on the movements of the blockading Anglo-French fleet.[1]

United States of America. As one would expect the main aim of telegraph links in the US was commercial. Boston had several systems, starting in 1801 and remaining, more or less continuously, in service until the advent of the electric telegraph in 1853. In New York the first telegraph was set up to watch the British Navy during the War of 1812 but fell into disrepair afterwards. There was a commercial line between New York and Philadelphia from 1840 to 1845. San Francisco had a commercial line for reporting the arrival of ships from 1849 to 1853. A notable feature of the line was the station on Telegraph Hill, which is still one of the most desirable areas of San Francisco, owing to its commanding view of the city and port.

[1] The telegraphs were nicely visible from the ships, offering ideal target practice for the bored crew. The semaphores were speedily repaired but the numerous holes in the buildings represented a health hazard.

Geostationary Orbits: Calculation of the Height of the Orbit above the Earth

The condition for keeping a satellite in stable orbit around the Earth is that two forces cancel each other. One of them is the gravitational force, F. It was found by Newton that two masses attract each other. The two masses in the present case are that of the Earth, M, and that of the satellite, m. This force is inwards, towards the centre of the Earth. It is equal to

$$F = gMm / R^2,$$

where g is the known gravitational constant, and R is the distance from the satellite to the centre of the Earth. The second force acting upon the satellite is the centrifugal force of circular motion that acts outwards,

$$f = 4p^2 mR / T^2,$$

where T is the period of the motion around the Earth. For a geostationary satellite T is equal to 24 hours. From the equality of the forces

$$gMm / R^2 = 4p^2 mR / T^2.$$

From the two equations above, using the rules of algebra, we obtain the distance from the centre of the Earth as

$$R = \text{cubic root}(gMT^2 / 4p^2).$$

Inserting into this equation the values of g, M, and T we find

$$R = 4.23 \times 10^7 \text{ metres.}$$

This still needs some correction because this is the distance of the satellite from the centre of the Earth. Subtracting the radius of the Earth we get the final result:

$$R = 3.59 \times 10^7 \text{ metres.}$$

To restate in kilometres: the geostationary orbit is 35,900 km above the Earth.

Bibliography

Aschoff, V. (1984, 1988). *Geschichte der Nachrichtentechnik* (2 vols). Springer Verlag, Berlin.

Bata, P. and Carré, P. A. (1985). *Presse, photographie et télécommunications de 1850 à 1940.* *Télécommunications*, **56**, 55–61.

Belloc, Alexis (1888). *Le télégraph historique.* Librairie de Firmin-Didot.

Bertho, Catherine (1981). *Télégraphes et téléphones de Valmy au microprocesseur.* Le Livre de poche.

Bertho, Catherine (1984). *Histoire de la télécommunications en France.* Éres, Toulouse.

Bloom, John (2016). *Eccentric Orbits: The Iridium Story.* Atlantic Monthly Press, New York.

Bray, John (1995). *The Communications Miracle.* Plenum Press, New York.

Bray, John (2002). *Innovations and the Communication Revolution.* IET, Stevenage, UK.

Bright, Charles (1898). *Submarine Telegraphs.* Crosby Lockwood and Son, London.

Bruce, Robert V. (1973). *Alexander Graham Bell and the Conquest of Solitude.* Gollantz, London.

Cairncross, Frances (2001). *The Death of Distance.* Orion Business Books, London.

Cochrane, Peter (1997). *Tips for Time Travellers.* Orion Business Books, London.

Costigan, D. M. (1971). *FAX: The Principles and Practice of Facsimile Communication.* Chilton, Philadelphia.

Dailey, Stephanie (1984). *Mari and Karana, Two Old Babylonian Cities.* Longman, London.

Feydy, Julien (1995). Le pantélégraphe de Caselli. *Musée des arts et métiers: La revue*, June 1995, 50–7.

Feyerabend, E. (1933). *Der Telegraph von Gauss und Weber im Werden der elektrischen Telegraphie.* Reichspostministerium, Berlin.

Fischer, Claude S. (1992). *America Calling: A Social History of the Telephone to 1940.* University of California Press, Berkeley.

Flatow, Ira (1992). *They All Laughed.* HarperCollins, New York.

Gabor, Luca (1993). *Telefon Hirmondo.* Magyar Radio, Budapest.

Hall, Brian N. (2017). *Communications and British Operations on the Western Front 1914-1918.* Cambridge University Press, Cambridge, UK.

Hawkins, Paul M. (2017). *Point to Point: A History of International Telecommunications during the Radio Years.* New Generation Publishing, London.

Headrick, Daniel R. (1991). *The Invisible Weapon.* Oxford University Press, New York.

Hecht, Jeff (1999). *City of Light.* Oxford University Press, New York.

Hunt, Bruce J. (1991). *The Maxwellians.* Cornell University Press, Ithaca, NY.

Huurdeman, Anton A. (2003). *The Worldwide History of Telecommunications.* Wiley, IEEE, New York.

Jolly, W. P. (1972). *Marconi.* Constable, London.

Keshav, S. (1997). *An Engineering Approach to Computer Networking.* Addison-Wesley, Reading, Mass.

Kieve, Jeffrey (1973). *The Electric Telegraph: A Social and Economic History.* David and Charles, Newton Abbot, UK.

Lebow, Irwin (1995). *Information highways and Byways.* Institution of Electronic and Electrical Engineers, New York.

Meyer, Hugo (1907). *Public Ownership and the Telephone in Great Britain.* MacMillan, New York.

Morris, P. R. (1990). *A History of the World Semiconductor Industry.* Peter Peregrinus, London.

Ornstein, Martha (1913). *The Role of Scientific Societies in the Seventeenth Century.* University of Chicago Press, Chicago.

Oxlade, Chris (2017). *The History of Ttelecommunications.* Heinemann, Portsmouth, NH.

Pierce, John R., and Michael, Noll A. (1990). *Signals: The Science of Telecommunications*. Scientific American Library, New York.

Reid, T. R. (1985). *Microchip, the Story of a Revolution and the Men Who Made It*. Pan Books, London.

Reuter, Michael. (1990). *Telekommunikation*. R.V. Decker-Verlag, Heidelberg.

Smaryl, O. L. (1984). New technology and the Soviet predicament. *Survey*, **28**, Spring 1984, 109.

Standage, Tom (1998). *The Victorian Internet*. Weidenfeld and Nicolson, London.

Storey, Graham (1951). *Reuter's Century*. Max Parrish, London.

Thompson, Sylvanus P. (1910). *The Life of William Thomson, Baron Kelvin of Largs*. MacMillan, London.

Vogelsang Ingo, and Mitchell, B. M. (1997). *Telecommunications Competition: The Last Ten Miles*. MIT Press, Cambridge, Mass.

Wilson, Geoffrey (1976). *The Old Telegraphs*. Philimore, London.

Yarrow, G., and Vickers, J. (1988). *Privatization and Economic Analysis*. MIT Press, Cambridge, Mass.

Young, Peter (1991). *Person to Person: The International Impact of the Telephone*. Granta Editions, Cambridge, UK.

Index